高等院校土建学科双语教材（中英文对照）

◆ 建筑学专业 ◆

BASICS

建筑模型

MODELBUILDING

[德] 亚历山大·谢林 编著

王又佳 金秋野 译

中国建筑工业出版社

著作权合同登记图字：01－2007－3333号

图书在版编目（CIP）数据

建筑模型/（德）谢林编著；王又佳，金秋野译．—北京：
中国建筑工业出版社，2011
高等院校土建学科双语教材（中英文对照）◆建筑学专业◆
ISBN 978－7－112－11601－0

Ⅰ．建… Ⅱ．①谢…②王…③金… Ⅲ．模型（建筑）－制作－
高等学校－教材－汉、英 Ⅳ．TU205

中国版本图书馆CIP数据核字（2009）第210932号

责任编辑：孙　炼
责任设计：郑秋菊
责任校对：关　健

高等院校土建学科双语教材（中英文对照）
◆ 建筑学专业 ◆
建筑模型
[德] 亚历山大·谢林　编著
王又佳　金秋野　译
*
中国建筑工业出版社出版、发行（北京西郊百万庄）
各地新华书店、建筑书店经销
北京嘉泰利德公司制版
北京建筑工业印刷厂印刷
*
开本：880×1230毫米　1/32　印张：4½　字数：145千字
2011年5月第一版　2011年5月第一次印刷
定价：**16.00**元
ISBN 978－7－112－11601－0
（20274）

中文部分目录

\\ 序　7

\\ 作为一种表现方法的建筑模型　82

\\ 模型的分类　84
- \\ 概念模型　85
- \\ 城市设计与景观模型，基地与地形　85
- \\ 建筑/房屋模型　89
- \\ 室内模型　91
- \\ 细部模型　92

\\ 设计与理念的发展　94
- \\ 色彩与材料　94
- \\ 构成与组成部分　95
- \\ 抽象与精细程度　96

\\ 设备、工具与技法　98
- \\ 切割　98
- \\ 粘结　102
- \\ 塑造、成型与浇注　103
- \\ 模型工作室中的机器　105
- \\ 电热丝切割机　110
- \\ 电脑雕刻机　112

\\ 材料　113
- \\ 纸、纸板与卡纸板　114
- \\ 木材与木质材料　118
- \\ 金属　122
- \\ 塑料　124
- \\ 涂料与清漆　128
- \\ 石灰、黏土与模型黏土　129
- \\ 配景：树、人与车　130

\\ 从图纸到模型——步骤与方法 133
\\ 一些初步的想法 133
\\ 底板 133
\\ 制作单独的建筑构件 135
\\ 组装构件 138
\\ 最后的工作与配景 138
\\ 表现 139
\\ 总结 140

\\ 附录 142
\\ 致谢 142
\\ 图片出处 142

CONTENTS

\\Foreword _9

\\The architectural model as a means of representation _10

\\Types of models _13
\\Conceptual models _13
\\Urban design and landscape models, site and topography _14
\\Architectural/building models _16
\\Interior models _19
\\Detailed models _21

\\Design and concept development _23
\\Colour and materials _23
\\Composition and proportion _24
\\Abstraction and level of detail _25

\\Equipment, tools and techniques _28
\\Cutting _28
\\Gluing _32
\\Modelling, shaping and casting _35
\\Machines in the modelling workshop _37
\\Hot wire cutters _43
\\Computer milling _43

\\Materials _46
\\Paper, paperboard and cardboard _47
\\Wood and wood-based materials _51
\\Metals _56
\\Plastics _59
\\Paints and varnishes _64
\\Plaster, clay and modelling clays _65
\\Accessories: trees, figures and cars _67

\\From drawing to model – steps and approaches _70
\\A few preliminary thoughts _70
\\The mounting board _70
\\Making individual building elements _72

\\Assembling the elements _75
\\Final tasks and accessories _75
\\Presentation _76

\\In conclusion _78

\\Appendix _79
\\Acknowledgements _79
\\Picture credits _80

序

模型是建筑设计的一种表达方式。因为它能够帮助设计师获得即将建成的建筑环境的空间印象，因此无论是在学习建筑学阶段还是在职业实践中它都是一种重要的表现手法。虽然三维的草图可以传达出空间的印象，但模型则允许观察者以自己的方式选择自己的视角去体验空间。

而且模型也是设计师的重要工具之一——它可以帮助设计师获得正确的比例与形式，同时也可以以三维的方式检验勾画出来的想法，并开始发展他们的理念。模型就是以这种具体的方式为设计与决定过程提供支持的。

这一系列基础教材旨在为第一次接触这一专业与学科的学生提供有益的与实用性的说明。它通过容易理解的介绍与实例来说明其中的内容。在每一章当中都系统地、深入地阐释与论述了其中最重要的原则。这一系列丛书不是编辑了专业知识的宽泛的纲要，而是意在对一个科目作最初步的介绍，并向读者传授必要的、可以熟练操作的专业技术。

这一卷探讨了以模型的形式表达缩小了比例的建筑。模型制作已经成为一种独立的艺术形式，有着自己的工具、技术和材料。因为学生们经常会不得不自己学习制作建筑模型的技术与通常的规则，本书将会介绍一些背景性的知识与实践性的技巧。

除了介绍不同种类的模型、常用的工具与机器设备，我们还将会系统地讨论适合的材料，并以其组合效果的视角来诠释它们。在本书中对于建筑模型制作过程的典型的描述还会包括一些提示与小贴士，它们可以帮助学生在模型制作所提供的多样性机会中受益，并将他们的设计转化成为富于美感与表现力的模型。

编者：贝尔特·比勒费尔德

FOREWORD

Models are a way of representing planned structures. Since they help create a spatial impression of what will become the constructed environment, they are an important means of presentation both while studying architecture and in professional practice. Although three-dimensional sketches can convey a spatial impression, models allow viewers to choose their own perspective and experience space in an individual way.

But models are also an important tool for designers – one that helps them to arrive at the right proportions and form, as well as to review sketched ideas in three dimensions and to develop their ideas in the first place. In this very concrete way, models provide support for the design and decision-making process.

This "Basics" book series aims to provide instructive and practical explanations for students who are approaching a subject or discipline for the very first time. It presents content with easily comprehensible introductions and examples. The most important principles are systematically elaborated and treated in depth in each volume. Instead of compiling an extensive compendium of specialist knowledge, the series aims to provide an initial introduction to a subject and give readers the necessary expertise for skilled implementation.

This volume examines scaled-down representations of buildings in the form of models. Modelmaking has become an independent art form with its own tools, techniques and materials. Since students are often forced to learn the techniques and general rules of model building on their own, this book will present background knowledge and practical tips.

In addition to introducing different types of models, common tools and machines, we will systematically discuss suitable materials and explain them with an eye toward their effects as compositions. The description of the typical model-building process includes tips and pointers that will enable students to benefit from the diverse opportunities of modelling and to transform their designs into aesthetic and representational models.

Bert Bielefeld
Herausgeber

THE ARCHITECTURAL MODEL AS A MEANS OF REPRESENTATION

What is modelmaking?

In the early Renaissance in Italy, modelmaking evolved into the most important means of architectural representation. It not only supplemented architectural drawings, but was often the primary method used to convey ideas and depict spaces. Ever since then, architects, engineers and clients have used models to represent designed buildings.

Plans (design sketches such as technical plans) and architectural models are both means of depicting buildings and spaces, yet plans convey only two dimensions. Once the design and the project have been thought out and sketches and details drawn, it is time to tackle the object in space. While it is true that only the finished building can communicate a complete understanding of three-dimensional effects, models anticipate the subsequent construction process. Seen this way, models represent architecture on a smaller scale. Work on models is of great importance, particularly in architectural studies, since students do not usually have the opportunity to build their own designs.

Motivation: why build models?

A model is not absolutely necessary to complete a design or architectural assignment successfully, yet it can be a useful tool in many ways: the model's scaled-down size makes it possible to examine the quality of the design and allows designers to develop a feel for space, aesthetics and materials. The additional advantages of a model include its communicative and persuasive potential: a model helps designers to demonstrate to themselves and others the quality of their ideas or projects. In addition, the model can serve as a control mechanism for assessing the building before it is built.

Working models – three-dimensional sketches

Architectural students are often confronted with the challenges of modelmaking in their first year at college. They quickly see that the models they built as children – trains, planes and ships – have nothing to do with the demands made on them in their studies. The green lawn from their train set is of no interest to professors or lecturers. The toys of childhood are now professional reality.

So how should students proceed? Let us assume they have been given a design assignment and require a model, on a pre-defined scale, that will play an important role in the way their proposed solution is evaluated. Whether they use a model in their design work will largely depend on both how they design and how the design evolves. If the spatial structure is

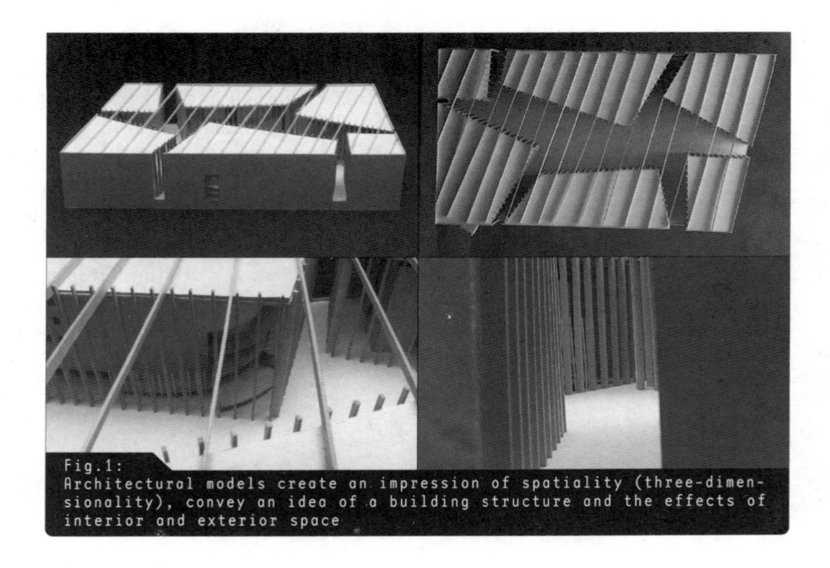

Fig.1:
Architectural models create an impression of spatiality (three-dimensionality), convey an idea of a building structure and the effects of interior and exterior space

of greater complexity than a two-dimensional drawing can represent, a model may be the only way to depict it. Simple models, so-called working models, enable designers to find solutions and test ideas. The ideas that do not work on the model can be rejected.

The working model accompanies the entire design process. Once this process is complete, the goal is then to illustrate the ideas and concepts that are essential for the design. The viewer must be convinced of the feasibility of both the concept and the proposed solution.

\\Hint:
Working models have earned their name because designers can "work" with them. They can, and should, be modified even if that harms their quality. It is therefore best to build a model that is easy to take apart and reassemble. Students can join parts with pins or easily removable rubber cement. Once two parts are glued together, they are committed to a particular form, which is not the purpose of a working model.

\\Tip:
Working models, which support the design process experimentally, can also be used effectively for presentational purposes if they are sufficiently refined. Elements such as building sections and topographic layers should be assembled temporarily both to ensure that the model can be modified and to avoid unnecessary stains and damage during work. Painting at a later stage in the process is an excellent way of enhancing a working model.

Presentation models – portraying and persuading

The presentation model, built with a great deal of effort to be almost perfect, marks the completion of the design process. At universities, this type of model is used to present a design idea – the concept. In architectural competitions, it depicts the proposed solution and competes with the designs of other participants. In both cases, the model supplements the architectural plans submitted. New media are often used to present drawings: three-dimensional computer-generated images provide a very realistic rendering of how the future structure will look. Photo-realistic representations are another modern way of simulating a spatial experience. A model cannot perform all these functions. It always remains an abstraction of the reality it portrays. Its only true function is to translate the sketched idea into three-dimensional form.

\\Tip:
It is always difficult to build a presentation model when under pressure to submit a design. It is often more efficient to make parts of the model in advance or use a site model as both working and presentation model. A presentation model does not need to be perfect or to be made of the very best materials. All that is needed is a very convincing model.

TYPES OF MODELS

Abstraction – the trick of miniaturization

A model is a more or less abstract, miniaturized representation of reality. But what does abstraction mean in modelmaking?

The opposite of abstract is concrete. In painting, "concrete" refers to an object that is portrayed as accurately as possible. In contrast, abstraction as applied to architectural models shifts the focus onto the subject matter, the informational value of the object portrayed and the spatial framework. What is at stake is not an accurate portrayal of reality but a process of simplification, which guides the eye to the model's essential features. It is crucial to find a suitable form of abstraction, one that reflects the selected scale. This point can be illustrated using the example of windows: on a scale of 1:200, a window is usually portrayed as a precisely cut-out aperture in the surface of the chosen material. On a scale of 1:50, a window is much easier to see. Glass is reproduced using a transparent material, and the window frame is built of small bars. Another example is façade cladding: on a very small scale, it cannot be depicted at all, but in larger models this element has greater relevance and is included in the design. The façade can be simulated very realistically using the right material.

›

Relationship between scale and representation

At the very start of a project, modellers must select the scale of their models. The dimensions in which an architectural object can be represented illustrate the role that scale plays. Depending on the scale and level of abstraction, there exist a number of model types, which will be explained below.

CONCEPTUAL MODELS (WITHOUT A CONCRETE SCALE)

A spatial pictogram can generally be described as a conceptual model. Here, the underlying idea of a design or a creative concept is depicted in an

Fig.2:
Windows shown at different scales

\\Example:
Try to reproduce the character and feel of surfaces in miniature. Rough or grainy surfaces can be simulated by sanding or "keying" cardboard and wood.

Structures can be simulated as follows:

_ Wooden cladding can be reproduced using small pieces of wood that imitate the real design.
_ Brick façades are simulated by cutting joints in the surface of the material.
_ Supports and beams can have the same shape as in reality.

entirely abstract manner as a three-dimensional object (e.g. using a metaphor as a basis). Material, form and colour highlight structures and create compositions. The model can, for instance, be used to visualize the results of urban space analyses at the start of the design process. By exploring a theme or place on a spatial yet abstract level, architects can alter or improve the view of that place. A model can support this approach.

URBAN DESIGN AND LANDSCAPE MODELS, SITE AND TOPOGRAPHY (SCALE 1:1000 / 1:500)

This type of model represents urban or natural environments. It is the first step in the representation process, since it shows the relationship

Fig.3:
Conceptual model

with the existing environment. In urban space, it is important to show how the context changes with the addition of a new structure.

This type of model is characterized by the highest level of abstraction. Buildings are reduced to "building blocks" – to abstract structures that reproduce building form and three-dimensionality in a highly simplified manner. Even so, the model includes the characteristic features of buildings such as recessions, projections, bay windows and roof designs. In its abstract form, the site – the scaled-down landscape – is simplified and depicted, in the chosen material, as a level plain. If the landscape slopes, it can be broken down into horizontal layers that are stacked on top of each other in the model.

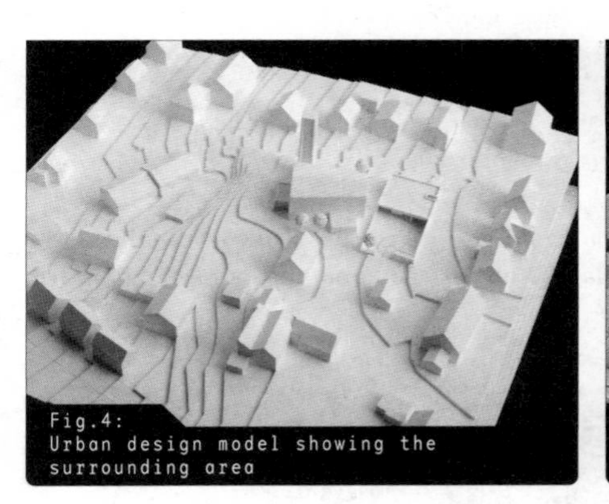

Fig.4:
Urban design model showing the surrounding area

Fig.5:
Urban design model of Berlin-Mitte

Fig.6:
Urban design model of Reykjavik showing the city and water

Raised-relief models

If the model is meant to reproduce uneven landscapes, the first step in building it is to conceive of the irregular natural terrain as a stack of horizontal layers. The more finely layered the material, the more precise and homogeneous the resulting model will be (width of the elevation layers: 1.0 mm or 2.0 mm). The work is based on a plan that shows contour lines or, at least, provides information on elevation. Once the real topographic situation is known, contour lines (curved, straight or polygonal) are drawn. Depending on the material, the modeller can cut out each layer with a utility knife or saw before arranging the layers on top of each other.

Insert models

A single assignment might include several designs, and urban design models are often constructed as "inserts" or group models to reduce the amount of work required to present them. Only one model of the surrounding area is made, and each participant is given a mounting board for the section on which he or she is working. This particular section is omitted from the urban design model so that the inserts can be interchanged.

ARCHITECTURAL/BUILDING MODELS (SCALE 1:200 / 1:100 / 1:50)

The building model is a common illustrative tool for simulating an architectural design. While larger objects such as museums, schools and

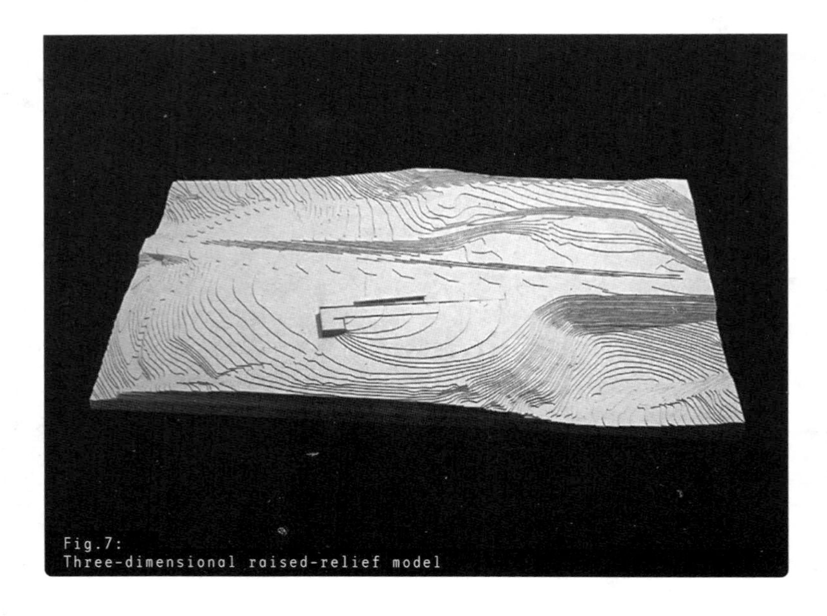

Fig.7:
Three-dimensional raised-relief model

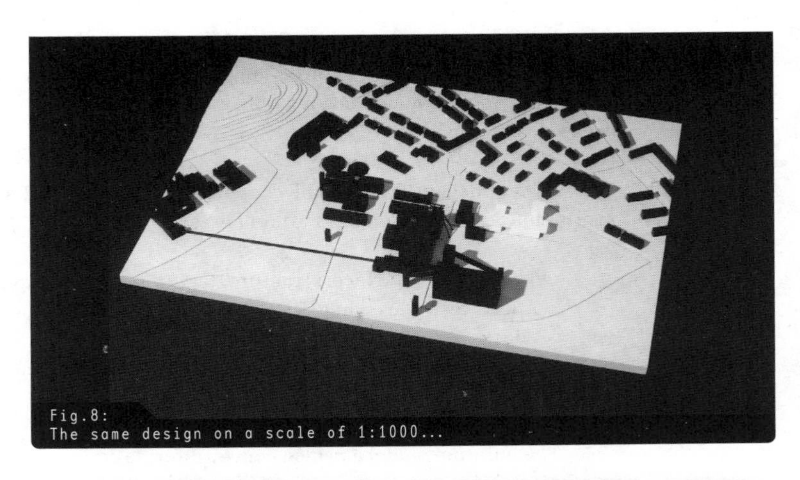

Fig.8:
The same design on a scale of 1:1000...

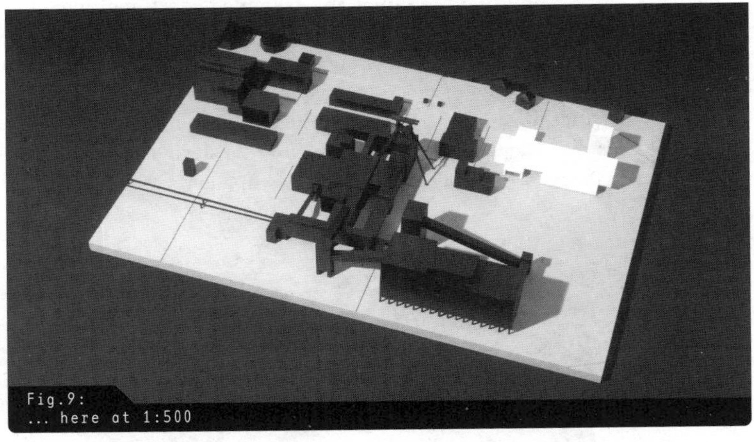

Fig.9:
... here at 1:500

Fig.10:
... and at 1:200

churches are usually represented on a scale of 1:200, 1:500 is common in competitions. In addition to three-dimensional forms and volumes, the design's many and diverse features play a more significant role here than in urban design models.

From drawing to model

Façade design is extremely important. On a reduced scale, the elements that are visible on the building exterior catch the eye, including:

_ The façade – surface, structure, features, material qualities
_ Apertures and what fills them (windows) – embrasure, wall widths
_ Roof type and design – specific details such as roof overhangs or parapets

The building model can also convey information on the interior space and building structure. For example, a sectional model (divided into two parts built on separate mounting boards) can provide views of important interior rooms. In combination with removable floors and other interior components, a modeller may alternatively use a removable roof element that affords a glimpse into the model from above.

Depending on the purpose of the design, the model can also be reduced to its structural or conceptual components for greater illustrative effect.

Fig.11:
Representation on different scales: 1:200 building model and 1:50 structural model

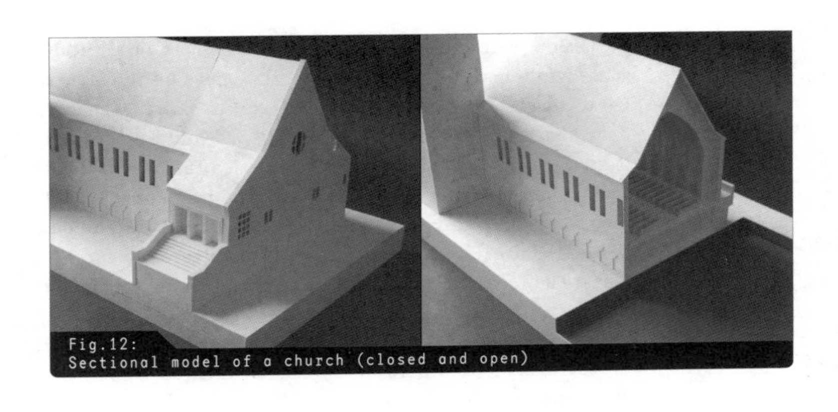

Fig.12:
Sectional model of a church (closed and open)

Fig.13:
Sectional model of a church (view of interior space)

INTERIOR MODELS (SCALE 1:20, 1:10, 1:5, 1:1)

When spaces such as bars, chapels or living areas are depicted, it is often advantageous to use a model to provide an accurate and detailed simulation of a real situation. Abstraction plays only a subordinate role in these models. The overriding goal is to portray real objects and materials on a small scale.

How can surfaces, elements and objects be reproduced on a small scale while maintaining the effect they have on the structure as a whole? Let us use the model of a chapel interior to explain the process. The design uses just a few materials for the finished building. Combined in a simple way, they possess the specific properties that are crucial for the spatial atmosphere and the overall concept. The linoleum floor of the real building is reproduced with linoleum in the model. Wood composite board is painted white to represent the real white plaster surfaces. Glass is used for glass. The resulting construction is both a three-dimensional model and a collage of materials.

The best way to simulate a real material is to use it in the model. For instance, model builders can reproduce fair-faced concrete in a large-scale model using concrete. They need only to build miniature formwork, fill it with a fine mixture of cement and sand, pack it down and let it dry. The results are impressive! In this case, modelmaking scales down the entire construction process, simulating not only the finished object, but the process itself.

Fig.14:
Model of chapel interior – scale 1:10

Fig.15:
Modelled interior of a tea salon - scale 1:20

The general rule is: the factors that have the greatest influence on the effect of the real building must be successfully reproduced in the model. This will make the model an effective representation of designed space. Modellers often achieve a stunning effect with their interior models, making it impossible for viewers to distinguish between a photograph of the modelled space and the real spatial situation.

DETAILED MODELS (SCALE 1:20, 1:10, 1:5, 1:1)

Detailed models are not only used in the field of interior design but also as structural or technical models (details). In principle, these models can be made on a scale of up to a 1:1, although in this case it would probably be more accurate to call them "prototypes". Model builders must first decide whether they wish to include furniture, lamps and similar details, and how they can do so most effectively. Creativity knows no bounds when it comes to putting old objects to new uses. Such objects can provide a basis for 1:10 scale models of tables and chairs or serve as simple building blocks or abstract volumes.

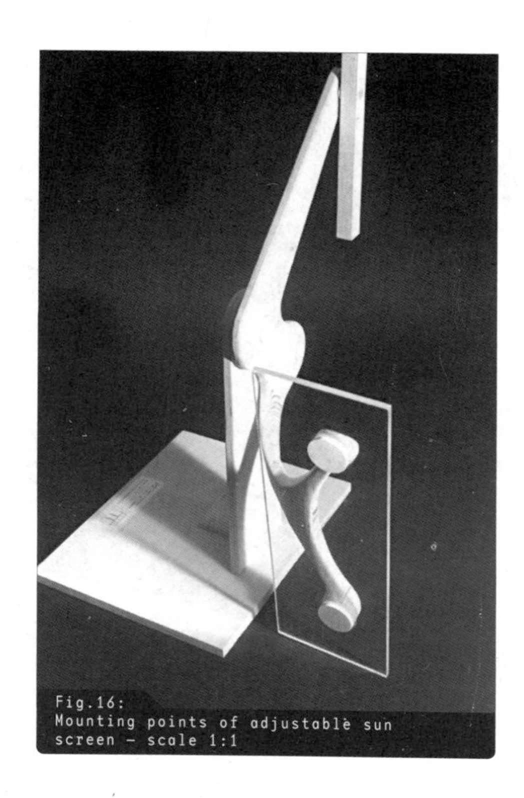

Fig.16:
Mounting points of adjustable sun screen - scale 1:1

Fig.17:
Multiplex plywood cabinet handle - scale 1:1

DESIGN AND CONCEPT DEVELOPMENT

Even a model needs a concept

Modelmaking is a creative process that requires a design concept in addition to the right tools and materials. › see chapter Equipment, tools and techniques The finished architectural model must be an aesthetic object that impresses people with its content and design. The following explanations are offered to help students develop designs:

- Colour – monochrome or polychrome models
- Material contrasts – differing material properties
- Composition
- Proportions
- Level of detail ("abstraction")

COLOUR AND MATERIALS

Monochrome models

After deciding whether a model should be built of wood, cardboard, metal or plastic, modellers must determine how many different materials they need in total. Generally speaking, one material will suffice if it can be modified in different ways. The advantage of surfaces with a uniform texture and colour is that the represented space will remain the focus and not compete with the material or the model itself for attention. Building monochrome models is a common approach. Most models in architectural competitions are executed as "white models" of plaster and polystyrene plastic: the goal is to direct the viewer's eye exclusively to the architectural or urban design project. Wooden models usually require only one type of wood. It goes without saying that the aesthetics of the selected material will be shown to best effect if other components do not detract from them. If it is nevertheless important to differentiate between components or elements, modellers can use paints, clear varnishes etc. › see chapter Materials

Differentiating elements and surfaces is part of the design concept. Accurately simulating the combination of elements in the model is thus a time-tested method of modelmaking.

A few architectural examples:

- Smooth plaster surfaces can be combined with a roughly textured brick wall.
- Solid concrete and stone elements can be used to contrast light, delicate structures of wood and steel.
- Transparent or translucent elements (glass building envelope) can be combined with opaque sections.

Fig.18:
A three-colour building model: the terrain is grey, the building anthracite black and natural brown (scale of complete model 1:200; detail 1:50)

If contrast is part of the architectural idea, it should generally be featured in the model. In such cases, representing materials in a simplified way to create a more abstract impression does not give the necessary clarity and precision. However, if different materials are combined, modellers should stick to what is absolutely essential to avoid a hodgepodge of materials.

COMPOSITION AND PROPORTION

Modelmaking replicates a process that has already taken place during the design work. The most important question is how materials can be combined to exploit their contrasts, whether hard or soft, dark or light, heavy or light, rough or fine. A great deal of experimentation may be necessary to make the final selection and to ensure that the composition of materials captures what the modeller wishes to convey. It is only after these issues have been resolved that the actual modelmaking should begin.

Interplay of elements

How do we "compose" a model? The goal is to enhance the design concept using the methods of modelmaking.

Modellers should first consider the following questions:

- How can the detail be selected so that the building model is in proportion to the dimensions of the overall representation?
- Will the project be positioned at the centre of the representation, or are there reasons to abandon this principle?
- What are the relationships between the individual elements within the model (colour, material qualities, proportions)?

›✎

\\Tip:
The size and format of the mounting board has a large influence on the effect of the model. You can hardly go wrong by choosing a square, the Golden Section, side proportions of 1:2, or by adapting the board to the format of architectural plans. Another option is to base the shape of the mounting board on that of the building: the effect of a long narrow building will be enhanced if a long, narrow mounting board is used.

ABSTRACTION AND LEVEL OF DETAIL

Along with the right materials, the mode of representation plays a key role in determining the final result. It is essential that all parts of the model have the same degree of abstraction. For example, there is no sense in accurately reproducing the site and the surrounding buildings if the model of the new building remains abstract.

Establishing the scale of the model will determine the level of abstraction. In modelmaking, abstraction means stripping things down to their essentials. Non-essential items can be ignored or left out. But what exactly are the essentials?

Abstraction = interpretative freedom

In this context, it is important to consider the effects of the selected level of abstraction: a precise, detailed model lays claim to portraying a sound, thought-out design. If the model is meant to provide a great deal of information, viewers will be given relatively little freedom in imagining it. A detailed, miniature version of the real structure will convey a very concrete idea of the intention of the design and keep viewers from imagining their own details. As a rule of thumb, a high level of detail is an advantage when dealing with clients and laypeople. The more realistic the model, the clearer the impression people will have of the building and the architecture.

What degrees of detail are possible in modelmaking? In principle, the only limits are what is technically possible, or what is possible within the given deadline. If a window is too small to be cut out with a utility knife at the selected scale, it should be left out.

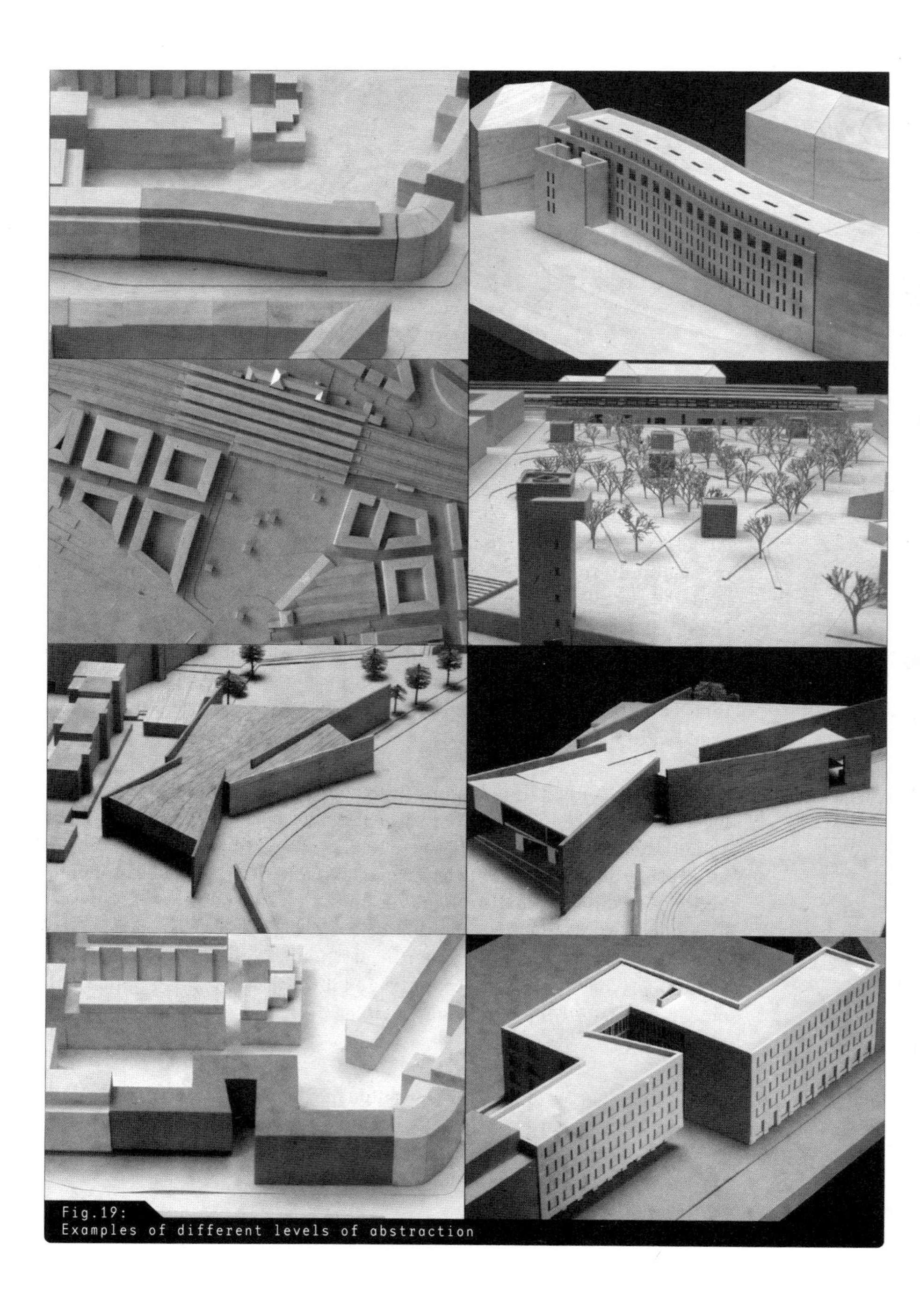

Fig.19:
Examples of different levels of abstraction

A more abstract model tends to convey principles and ideas, with details fleshed out at a later stage. The model remains conceptual. Architects often prefer this type of minimalist representation to grant free rein to the imagination and to enable various interpretations of the subsequent building. They are not forced to commit themselves.

Accessories

The use of trees, figures, cars and other accessories in modelmaking must also be considered in conjunction with the level of abstraction. It is often easier to convey a concrete idea of a design to a lay audience if the model depicts not only the architecture and the site but also elements that are familiar to them from daily life, such as cars, trees, plants and people. "Human scale" is the most important yardstick in real life, but it is not always apparent in a model. › see chapters Materials, Accessories: trees, figures and cars

Precision

Modelmaking is often associated with terms like perfectionism, exact workmanship and precise execution. These standards are generally justified when presentation models are made, but precision on its own does guarantee a good model. It is not even the prerequisite for one. An idea can be portrayed clearly without a lot of work. More important than perfection is the modeller's "signature" style, which we have mentioned above, and which gives the model the required expressive power. As a creative approach, a broad stroke can be as effective as the detailed representation of the large object on a miniature scale.

EQUIPMENT, TOOLS AND TECHNIQUES

When building small spatial objects, many students will feel the same passion that drew them to architecture as a profession in the first place. Others will prefer pencils and drawings to the tools of the modellers' trade. Whatever their preference, all architectural students will be confronted with the demands of modelmaking in their very first semester. What is the simplest way to build a model?

CUTTING

The simplest method is to work with a utility knife and a piece of paperboard or cardboard. The utility knife is an essential modelmaking tool since it can be used to cut a variety of materials in their original form. It is simple and inexpensive, and many different versions are available for different cutting tasks. We recommend purchasing a high-quality knife instead of the simple carpet knives found in many home improvement stores. The blade must be firmly attached and must not move during cutting. Ideally, the knife will fit comfortably in the hand so that it can be held and guided effectively. The selection of a knife is just as important as the choice of a good pencil.

In addition to the universally deployable utility knife, a variety of other cutting tools are available at speciality shops. The scalpel, the well-known surgical instrument, can be used to make fine cuts in materials or cut out very small windows from cardboard.

\\Hint:
Many modellers have had painful experiences with the utility knife, especially when working under time pressure. To minimize the risk of injury, you should always handle tools with the utmost care and only use them for the work for which they are intended. For instance, you should not to try to cut a hard piece of wood with a utility knife due to the danger of slipping.

\\Tip:
When using a utility knife to cut cardboard, you should hold the knife as close to the surface as possible in order to make precise, clean cuts. If this is not done correctly, the material may tear and increase the risk of injury since the sharp blade is more likely to slip.

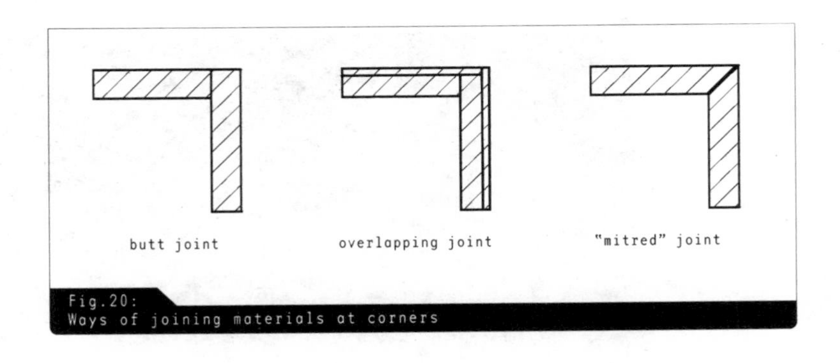

Fig.20:
Ways of joining materials at corners

Fig.21:
Corners – examples and approaches

How should cuts be made? First, the required edge should be marked with a thin pencil; after cutting, this thin mark will no longer be visible. The edge should then be cut at a 90° angle to the surface so that the individual parts can be butt-jointed or glued together.

At corners the edges of the material can also be cut at a 45° angle to form mitred joints. Special utility knives with slanted blades are available to make such cuts. Another method is to make a wooden board with a 45°

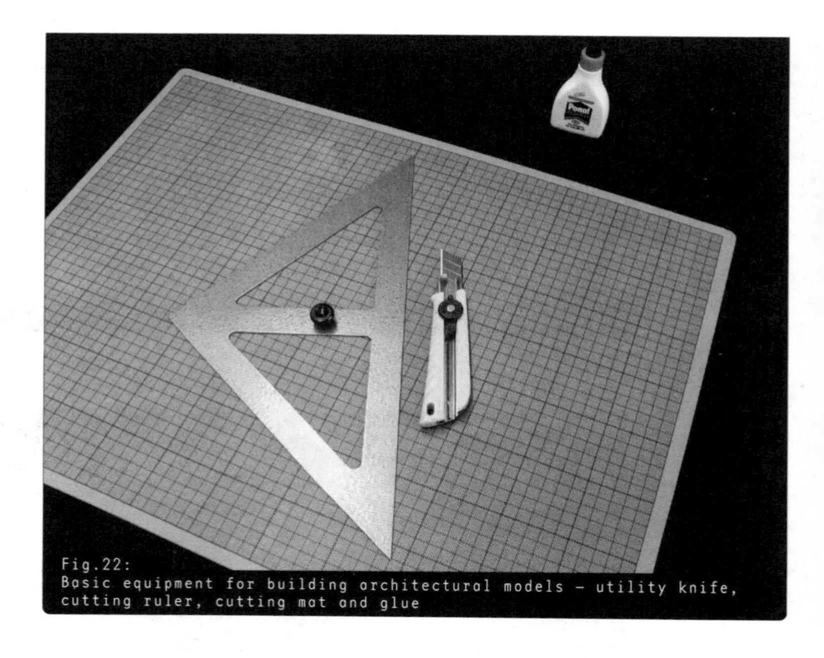
Fig.22:
Basic equipment for building architectural models – utility knife, cutting ruler, cutting mat and glue

edge and use it as a kind of a template for guiding the blade of a standard utility knife.

In other cases, a metal-edged ruler can be used to guide the knife and ensure a straight cut. Another important tool is a hard cutting mat, which not only protects the tabletop but also makes for a better quality cut than a soft base. We recommend plastic cutting mats.

These are the only tools a modellers needs to build simple, practical working models out of cardboard and pasteboard. But other tools are necessary to intensively work the material, including:

- Sandpaper: required to smooth surfaces, finish cut edges and remove sharp pieces from cut-out holes. If sandpaper is wrapped around a firm, stable base such as a sanding block, it is easier to move back and forth over the part that needs sanding.
- File: excellently suited for finishing the corners and edges of a variety of materials. Files are available for both wood and metal. It is

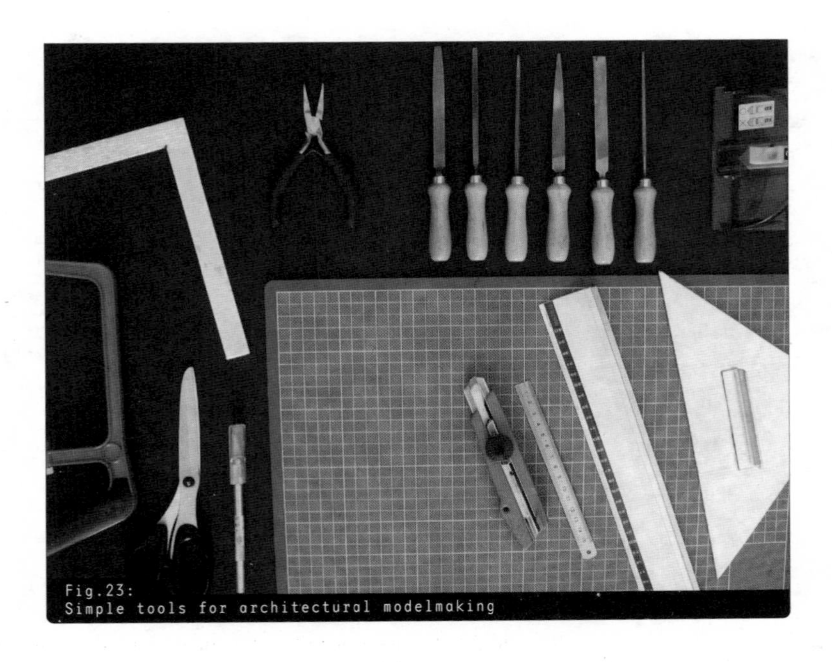
Fig.23:
Simple tools for architectural modelmaking

generally possible to use the same tools when working on plastics or metals – for instance, a fine iron saw lends itself well to cutting plastic.

_ Tweezers: increase the precision of the human hand and allow modellers to work on even the smallest parts. Small-scale models may contain parts that are only a few millimetres in size and nearly impossible to grasp with the fingers.
_ Measuring instruments ("sliding calipers"): are used to measure diameters and cross sections accurately. Since the trained eye can detect even the tiniest size differences on a small-scale model, precise measurements are an important aspect of modelmaking. Precision extends to one tenth of one millimetre, a unit that cannot be measured with normal rulers.
_ Measuring tools (ruler and measuring stick): for the precise measurement of lengths and dimensions.

✎
\\Tip:
Before gluing the model together, you should first find out about the most suitable glue. Most glues are sold in containers and do not include a tool for applying the glue properly. As a result, you may apply too much glue and leave unsightly stains.

One way to remedy this situation is to buy a standard syringe with a wide canula from a pharmacy and fill it with the selected glue. This method can be very useful when gluing transparent plastic films since stains on these materials are eye-catching and bothersome.

An even simpler method is to apply daubs of glue using a toothbrush or a thin wooden stick.

GLUING

Depending on the materials to be glued, modellers have a wide range of products at their disposal. Whereas all-purpose glues can be used universally for many different materials, white glue is the perfect choice for all types of wood, wood-based materials and cardboard. ›✎

Table 1:
Overview of glues

Glue	Properties	Application
All-purpose glue	Generally a solvent-based artificial resin adhesive that is transparent, viscous and may become stringy when applied. Reacts with some plastics (e.g. Styrofoam), melting their surfaces. Usually dries in just a few minutes. A light irritant; non-aging	Can be used for various materials (e.g. cardboard, wood, plastics, metals, glass, fabric etc.). Parts can be glued to the same or different materials. The glue does not cause cardboard to buckle and is often all that is needed in modelmaking thanks to its great diversity.

White glue (wood glue)	Has a white, viscous consistency and dries by absorbing the moisture in the bonded material. Transparent and slow drying, meaning the glued surfaces can be moved for a while after application. Fast-drying "express glue" dries in 3–5 minutes.	Ideal for woods, wood-based materials, cardboard and paperboard. The parts must be pressed together for an effective bond. The glue's high moisture content may change the shape of the material and cause cardboard to buckle. Not suitable for plastics, metals or materials that cannot absorb moisture.
Contact glue	Used to bond large surfaces. The glue is applied to both parts and bonds with itself. After application, the parts must dry for a few minutes before they are bonded together, and it is important to apply firm pressure. Contact glue should only be used in well-ventilated rooms.	Ideal for gluing large surfaces (e.g. cardboard) in raised-relief models. Since the glue is applied to both parts and the room must be ventilated, this is a time-consuming method, but it has the advantage of not buckling the materials. Can be used for wood, cardboard, a large number of plastics, metals and ceramics.
Plastic-bonding adhesives	A fluid, usually clear glue containing solvent. Specially designed for plastics and the single-sided application of glue. The parts must be joined quickly while the glue is still wet, and the surfaces must be free of dust and grease.	Can be used for many thermoplastics, including polystyrene, PVC and perspex, but not suitable for polyethylene or polypropylene. Can also be used to bond wood or cardboard (see all-purpose glue), and are more effective than all-purpose glue for plastics.
Super glue	Transparent, very fast-drying glue; viscous and non-dripping.	Ideal for connections that cannot be held together and that require an instant bond.

Spray glue	Colourless, UV-resistant, CFC-free glue that comes in a spray bottle and does not change colour when applied. Due to its low moisture content, it does not permeate the material. Spray glue should only be used out of doors or in well-ventilated rooms.	Ideal for application to large surfaces, such as in cardboard raised-relief models. The materials may buckle or warp slightly, but do not usually wrinkle. Can also be used to glue paper or cardboard to different backgrounds, especially when large surfaces must be glued.
Solvent	Used to bond plastics. The solvent melts the surface of the material, and pressure is applied to "weld" the two parts together. Example: dichloromethane (methylene dichloride), which like all organic solvents is extremely harmful to human health.	For gluing polystyrene, perspex or polycarbonates; bonds both parts without residue by dissolving the thermoplastic resin. Use only in very well-ventilated rooms.
Rubber cement	An elastic adhesive that is easy to remove, normally without any residue. If applied to both parts, it creates a permanent bond.	Can be used for various materials including paper, cardboard and plastics. Ideal for use in working models or montages
Double-sided tape	A basic alternative to fluid bonding agents, useful because of its instant bond. The properties of the bonded materials are not adversely affected by moisture.	Can be used for bonding large surfaces, and ideal for all types of materials, including PE and PP, which cannot be joined with other adhesives. The materials do not buckle, warp or wrinkle. Not suitable for joining small points.

MODELLING, SHAPING AND CASTING

Working with modelling clay

An alternative method for building working models is to use plasticine or another modelling clay.

Plasticine is usually sold as a greyish-green modelling material whose plasticity varies with temperature. It is hard to knead plasticine at room temperature, but it becomes softer and easier to work once it is slightly warmed up (e.g. on a warm radiator). Plasticine can also be heated in a pot, and in fluid form, it can be applied to a surface with a putty knife or brush.

Modelling clay lends itself well to experimental work with a model. It is excellently suited to studies of urban space and enables modellers quickly to build many different versions of the same object. Small structures can be quickly and easily cut off with a knife.

By working the clay with their hands or tools, modellers can gain a feel for the manual work and its results. The finished object may not be a very precise model, but the modelled form can be a very expressive way of representing space.

Plaster and pouring compounds

Plaster (gypsum) is an inexpensive material that is very easy to work with. Even so, making a plaster model is a time-consuming process because it requires two steps:

_ First, an exact negative mould is made of the subsequent three-dimensional object and liquid gypsum is poured into it. The precision of the mould determines the quality of the final result.
_ After the plaster dries, the cast object is removed from the mould.

Plaster models are often used in architectural competitions, particularly for urban design models, since here the same site model must be

Fig.24:
Plasticine models

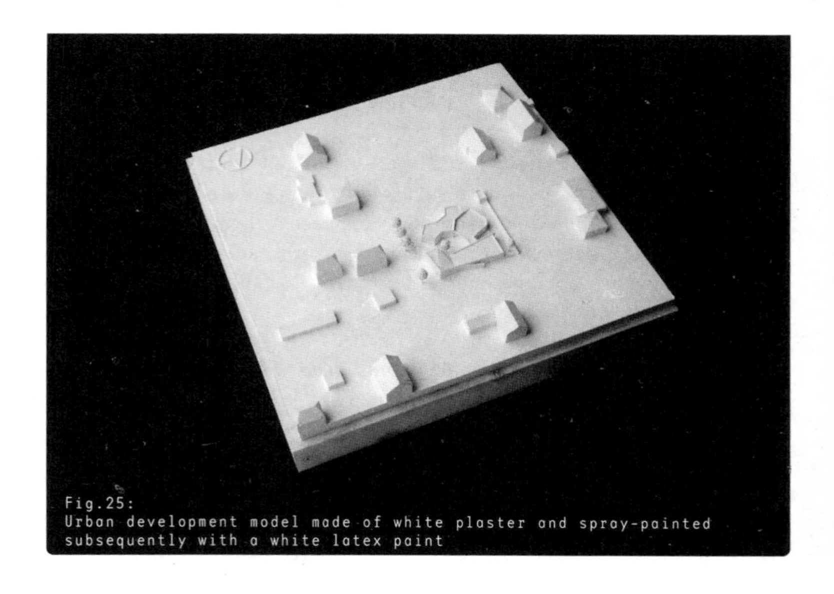

Fig.25:
Urban development model made of white plaster and spray-painted subsequently with a white latex paint

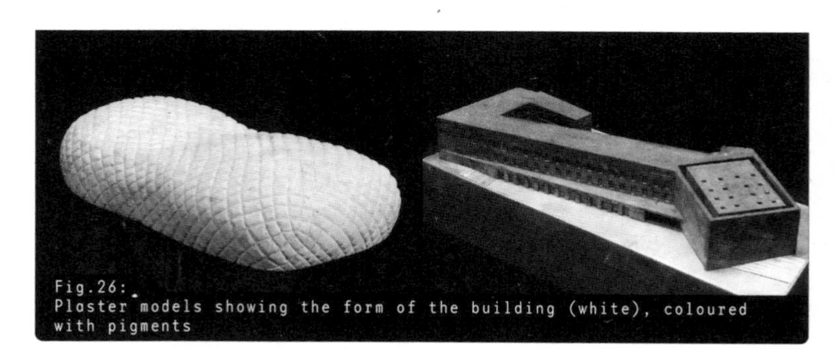

Fig.26:
Plaster models showing the form of the building (white), coloured with pigments

Fig.27:
Concrete model portraying a solid concrete building

made several times and using a single mould is convenient. Colours can be portrayed by adding pigments or liquid paint to the plaster or by smoothing and painting the surface after it dries. These models not only create a solid, heavy impression, but are solid and heavy themselves.

It goes without saying that other pouring compounds can be used besides plaster. If fair-faced concrete needs to be represented, modellers can use concrete in the model.

MACHINES IN THE MODELLING WORKSHOP

Working with machines

The simple tools and methods described here are sufficient for working with easily processed materials. However, many materials require the use of machines and professional tools in order to achieve the desired result.

Tools in the carpenter's workshop

Many professional architectural modellers use the same tools that can be found in most carpenters' workshops:

_ Handsaws
_ Files, rasps and sanding blocks
_ Planes
_ Chisels and mallets
_ Set squares

Sawing

A power saw is commonly used when a utility knife is no longer sufficient. Although a knife can be used to work on veneer, wood is a material that usually requires the use of a saw.

In addition to the usual commercial table saws – saws with round blades and benches – table saws of a smaller size, called micro table saws,

Fig:28:
Typical handheld tools in a wood workshop

Fig.29:
Circular saw

Fig.30:
Jig saw

are also available for (professional) modellers. If the correct blade is used, these fine power tools can be used for cross cutting and profiling wood, as well as for cutting plastic. It is very helpful to have precisely these kinds of tools at hand when building wooden models. With the use of the fence, it is possible to cut materials both across and with the grain. It should be noted, however, that students will require a certain amount of training and time to become familiar with such saws and the specific features of materials to be worked.

›✎

›✎

In addition to the table saw, which is useful for straight cuts, the band saw is an important tool for curved and free-form cuts. This machine is useful for sawing solid wood cross sections. Jig saws are also useful for cutting free-form and curved lines.

Planing

A plane is used to reduce the thickness of wood or to modify the surface of cross sections. By planing the surface of terrain models, the modeller can make differences in elevation more apparent.

✎
\\Tip:
Using wood for architectural modelling is extremely attractive, as the result lives from the aesthetic of this material. Even so, use of the required machines like the table saw or grinding machines should be supervised by professional joiners or modellers, particularly at the beginning.

✎
\\Tip:
In addition to tools available in hardware stores and DIY centres, creative modellers can often make their own tools using very simple materials. A good example is a miniature file consisting of a thin piece of square iron pipe and fine sandpaper. It is also possible to use everyday items. For instance, a clothespeg works well as a clamp when gluing.

Fig.31:
Band saw

Fig.32:
Table saw

Fig.33:
Micro table saw for precise cutting

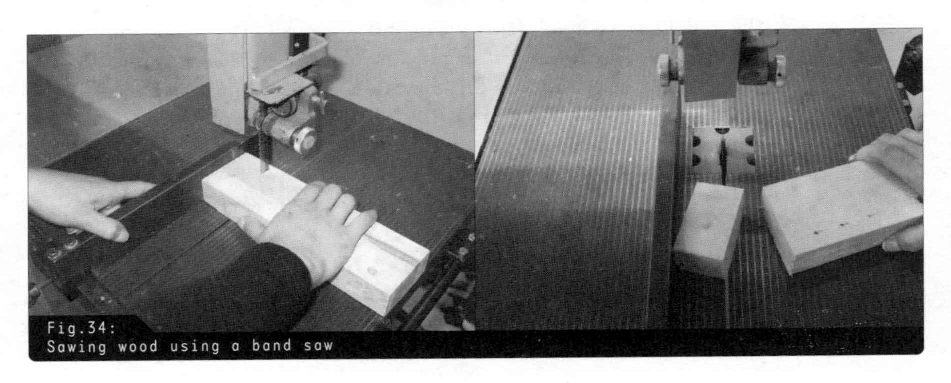
Fig.34:
Sawing wood using a band saw

Fig.35:
Planing a surface using either a traditional smoothing plane or an electric one

Fig.36:
A handheld power drill with no type of guidance system

Drilling

A drill is useful for making connections, for example when dowels are used on the base of a model to represent columns or trees. Holes bored in the base material are perfect connections for these elements and increase stability as well.

Of utmost importance when using a drill is the correct guidance of the tool. In workshops, drills are usually clamped into a drill stand so that the angle at which the drill enters the material can be set precisely and locked in place. The micro drills available for modelling have very small chucks and can be used with drills with a diameter of less than one millimetre.

Milling

A milling machine enables a modeller to cut into the surface of wood.

The different types of milling bits › see Fig. 37 that can be attached to the machine allow modellers to cut roads, rivers, depressions and other features into the surface of a raised-relief model. This machine offers new

Fig.37:
A drill in a drill stand – for precise and right-angled drilling

Fig.38:
A milling machine

woodworking options in addition to those possible with the saws outlined above.

Sanding

There are various ways to sand an object in order to produce the final shapes and surfaces. Power tools make the work much easier than manual sanding with sandpaper. It is of fundamental importance when sanding that the proper type of sandpaper – e.g. for metal or wood – be used for each material. Coarse sanding can be done using coarsely grained sandpaper, and a finely grained paper should be used afterwards. A power disc sander is a very helpful tool. A rotating sanding disc is combined with a flat work surface so that large surface areas can be finely sanded. Just as with table saws, micro sanders are also available to carry out the fine and precise operations necessary for precise modelmaking.

In addition to flat disc sanders there are also tools with rounded sanding surfaces, which are useful for the smoothing off round shapes. They include belt sanders and oscillating sanders, which are guided by hand across the surface in question.

Fig.39:
The fine sanding of wood with a disc sander, here for an urban design model

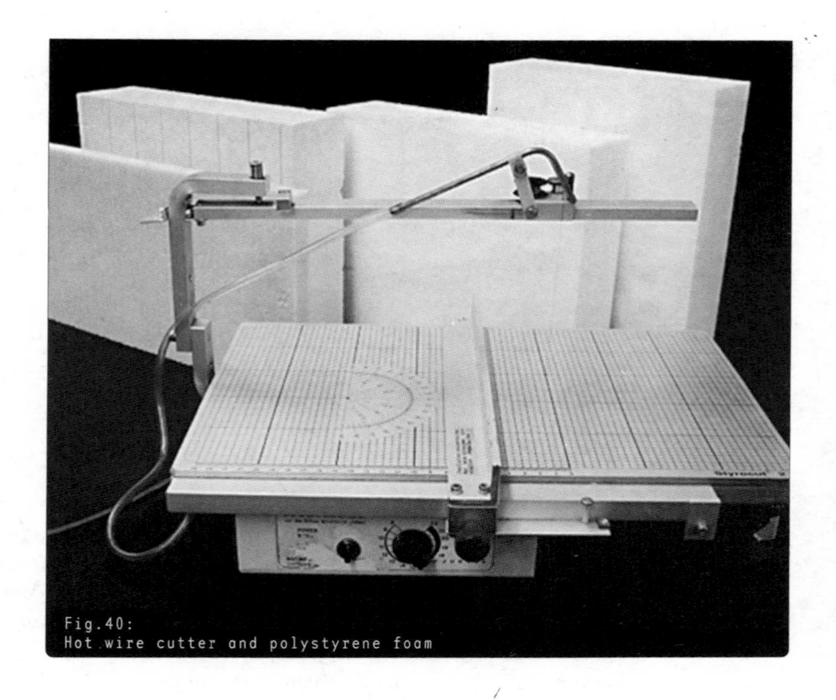

Fig.40:
Hot wire cutter and polystyrene foam

HOT WIRE CUTTERS

A hot wire cutter (Styrocut) is a common tool among architecture students. It allows for the simple, fast and precise cutting of thick polystyrene foam and can be used to make individual pieces for an urban design model or mass models during the planning process. The hot wire cutter consists of a cutting table and a fine wire, which is heated up by a low voltage energy supply and slices through the foam material.

COMPUTER MILLING

The use of computers has transformed the field of architecture, particularly in representing complex organic forms and structures. CAD software and CNC (computerized numerical control) milling machines – a modern digital tool for architectural modelling – have played the biggest role in this change. Strictly speaking, the use of CNC milling for three-dimensional representation involves nothing more than the use of CAD software for two-dimensional drafting. Just as the draughtsperson's methods have been digitalized, so has architectural modelling by the CNC milling machine.

\\Hint:
Depending on the design, CNC milling machines can speed up the modelling process enormously, particularly when reproducing façades with many apertures. Colleges, professional architectural modellers and companies that make moulds and prototypes commonly offer the use of their machines at fixed hourly rates.

Technically, CNC milling machines, which are usually controlled along three spatial axes (x, y and z) by the computer, are able to produce models with an extremely high degree of precision and perfection. The cutting and engraving bit cuts out geometrically and spatially complex shapes from a variety of suitable materials, just as a digital plotter can produce finer and more accurate line drawings than can be made using manual methods. In addition, models can be produced as often as desired, and

Fig.41:
CNC milling machine

\\Tip:
During the milling process, which involves digitally converting the two-dimensional drawing into a three-dimensional spatial object, all mistakes in the drawing will become visible. It is thus important to consider the following points:
_ Lines must always be cleanly and precisely connected at their endpoints.
_ Double lines cannot be used.
_ Lines must be assigned colours in such a way that the milling machine is able to define the right and left sides of a line.

the individual parts are always identical. This machine therefore opens a broad field of possibilities for graphic representation that were not available before. Some passionate modellers nevertheless feel that items produced using CNC milling machines have no character of their own.

Many universities now offer these new digital methods in addition to traditional modelling workshops as part of architectural studies. Depending on the type of machine, all necessary modelling data (base topography or information about a building, including the structure of the façade in vectorized form) must be available before any milling can take place. These data enable the machine to cut the correct shapes out of the material. Problems occur if the milling machine's software cannot adequately translate the CAD file (due to older file standards, for example). This means that everything has to be drawn once again.

To achieve optimum results with CNC milling machines, modellers must allow sufficient time for test runs with this modern technology. It is important to note whether the cutting head moves to the left or right of a pre-defined line, or directly on the line itself.

Not all materials are suited for use with a CNC machine. Aluminium, brass, steel and stainless steel can be used in sheet form (a thickness of 3 mm to 5 mm is common). Perspex of superior (moulded) quality can be used in thicknesses of up to 10 cm, as is the case for plastics such as polystyrene. Wood products such as plywood (e.g. birch and multiplex plywood, and MDF) can also be used.

MATERIALS

Which material for which model?

Strictly speaking, a model is nothing more than a scaled-down version of an existing situation or a future building. In the practice of modelling this means that actual surfaces are transformed into a model through a process of abstraction. An attempt is nevertheless made to retain the specific qualities of materials and the effect they might have, because in the end a building's appearance comes primarily from the sum of the materials from which it is made. Materials are matte or shiny, rough or smooth, heavy and solid or light and filigree. This raises the question of which materials the model should incorporate in order to simulate reality best. In this regard modelling is the same as real construction. A wide variety of materials are available. Some have been used in architectural modelling for a long time and are considered "classic" materials; others are new; and others still were originally intended for different purposes. Specific characteristics are necessary to create particular effects. Glass can be simulated using transparent materials, water by reflective materials, and walls can be reproduced using layered materials made of small modules with a regularly or irregularly structured surface.

Below, materials are described together with their uses and methods of preparation. Materials can be categorized on the basis of their specific sources:

_ Paper, paperboard and cardboard
_ Wood and wood-based products
_ Metals
_ Plastics

Other important materials are:

_ Paints and varnishes
_ Plaster and clay

\\Hint:
Most projects can be easily built using these materials. In addition their ease of use, they have the advantage of being inexpensive and readily available in many places.

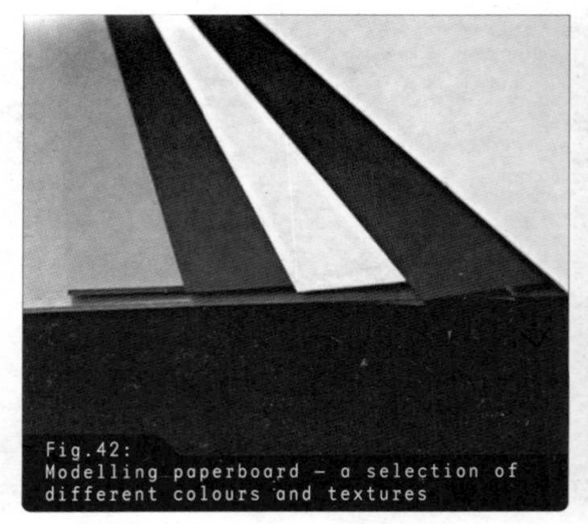
Fig.42:
Modelling paperboard – a selection of different colours and textures

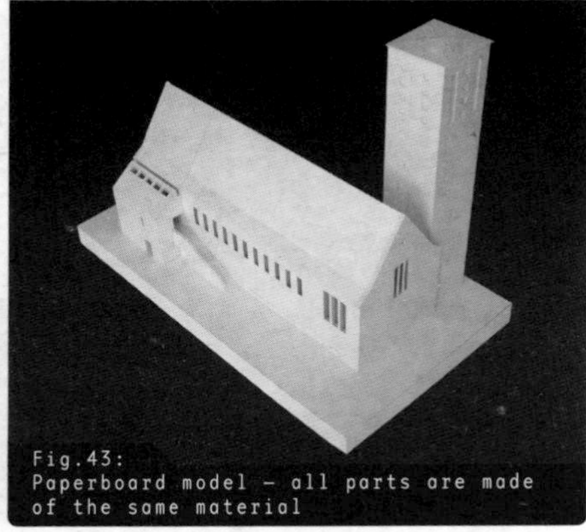
Fig.43:
Paperboard model – all parts are made of the same material

_ Plasticine and modelling clay
_ Prefabricated parts for modelling such as figures, vehicles and other accessories

PAPER, PAPERBOARD AND CARDBOARD

A wide variety of these materials exist on the market, and there is no limit to the creativity with which they can be used in architectural modelling. Many products, including grey paperboard, were actually developed for the packaging industry but are nonetheless very popular among modellers.

›

Table 2 (page 47) gives an overview of the most common types of paperboard. Paper, including coloured paper or tracing paper, is also a useful material for modelling. Cardboard, sold under names such as Mill blank, coloured board, glossy board or Chromalux board, can also be used, as can paper and cardboard with structured surface textures.

Photo cardboard

Photo cardboard, like thinner coloured paper, is particularly well suited for representing different coloured surfaces. It is available from many manufacturers in a wide variety of colours, and is a stable material for the work due to its thickness.

Glossy cardboard

Glossy cardboard's name leaves little doubt as to the material's special feature. Because of its glossy exterior this cardboard is always suitable for representing reflecting or mirrored surfaces. Its many different colours are also an advantage in addition to its shine.

Fig.44:
Model made of grey paperboard

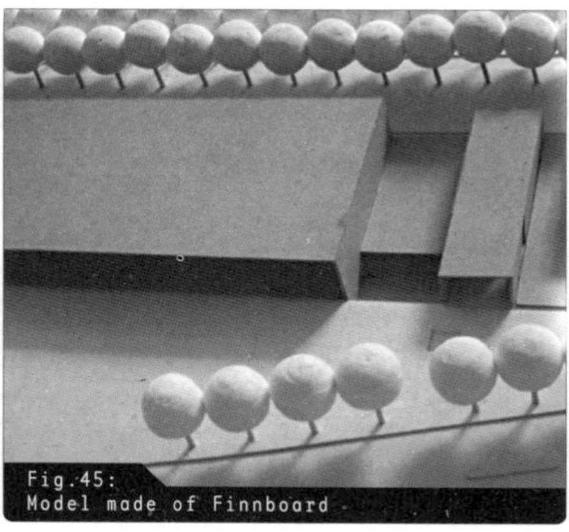
Fig.45:
Model made of Finnboard

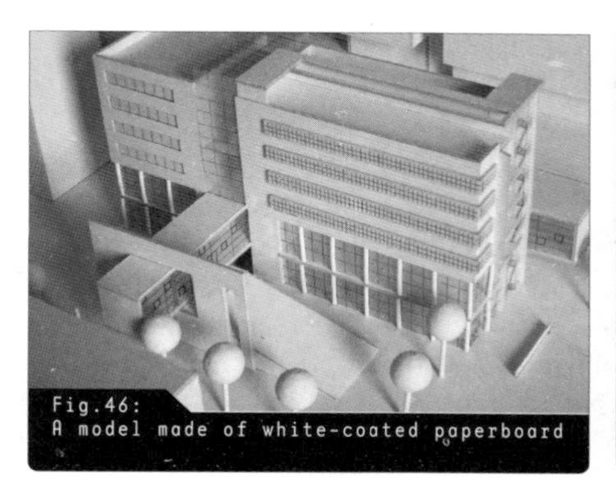
Fig.46:
A model made of white-coated paperboard

Fig.47:
A model made of dyed white paperboard (Chromalux)

Corrugated cardboard/ wellboard

Materials that are inexpensive or available for free are always very practical for building working models. Corrugated cardboard is primarily produced and used for packaging and can be "recycled" and used for modelling. The advantages that the material brings to packaging also apply to modelling:

_ It is easy to cut using a utility knife
_ Thicknesses of up to 6 mm make it easy to work with
_ The structure of this material (corrugated cardboard core, covered on both sides with smooth cardboard) means that it keeps its shape and remains stiff despite its light weight
_ It is also suitable for use with raised-relief models

Table 2:
Paperboard – uses and characteristics

Material	**Properties**	**Use**	**Processing**
Finnish wood pulp board (also known as Finnboard)	Made of wood fibres; beige, woody colour; darkens when exposed to sunlight (yellowing); smooth and rough textures; available in thicknesses approx. 1.0–4.0 mm	General use for topographic layers and buildings, interior models with simulated daylight (due to its light surface)	Easy to cut; easy to stick together using white or all-purpose glue; surfaces can be varnished or painted to prevent yellowing or aging
Grey paperboard	100% recycled; warm grey tone; smooth and rough textures; available in approx. 0.5–4.0 mm thicknesses according to manufacturer	General use for topographic layers and buildings	Easy to cut; easy to stick together using white or all-purpose glue; surfaces can be varnished or painted
Serigraphy board	Wood fibreboard with bonded texture; available in approx. 1.0–3.0 mm thickness	Ideal for spatial models (scale 1:50); simulation of light due to white surface colour	Easy to cut; easy to stick together using white or all-purpose glue; surfaces can be painted with acrylic paints or printed with silk-screen dyes

Fig.48:
Model made of corrugated cardboard showing a building extending below a site's surface (layers of earth)

Fig.49:
Combination of grey board (site) and birch plywood

Fig.50:
Urban design model made of one kind of wood (homogenous)

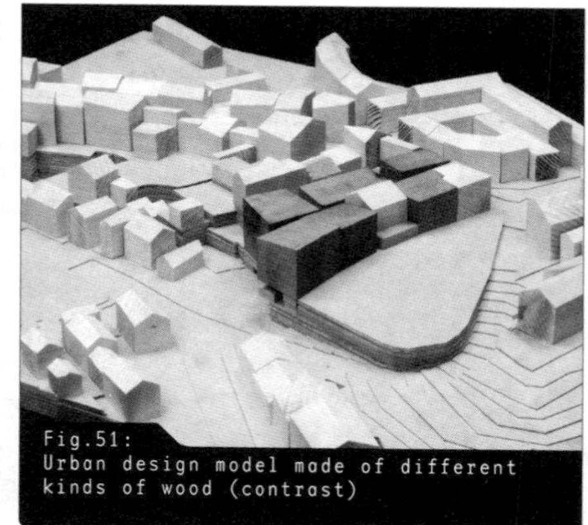

Fig.51:
Urban design model made of different kinds of wood (contrast)

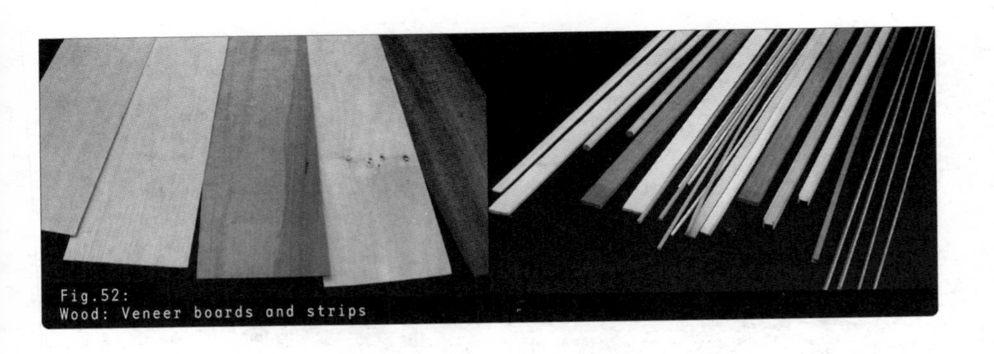
Fig.52:
Wood: Veneer boards and strips

WOOD AND WOOD-BASED MATERIALS

Wood has been used longer than any other "construction material" for building architectural models. Even Michelangelo used the wood of a linden tree to make a model of Saint Peter's dome. Wood is much more work-intensive than paperboard; the results, however, are much more impressive. Wood is used above all for presentation models. As a natural raw material, wood has its own aesthetic, which is expressed independently of an architecture model's shape, design and finish. Nuances in colour and the structure of the grain give wood the appearance of a "living" component of a model.

There are two basic categories: naturally grown and dried wood, and industrially processed wood-based products. Both are used in modelling.

\\Tip:
A wide variety of good-quality wood is available at timber yards as well as in small carpentry workshops and wood-processing firms. An overview of the various types of wood from all over the world is nearly impossible, but professionals are usually happy to advise architects and students about the correct choice of wood for modelling. It may also be possible to observe carpenters and cabinetmakers as they work, and learn about how they deal with wood.

Fig.53:
Rich in contrast, with an even grain (e.g. elm)

Fig.54:
Dark and finely grained (e.g. wengé)

Fig.55:
Grainy, yellow-beige wood (e.g. oak)

Fig.56:
Strongly grained wood, "busy" surface (e.g. zebrawood or Macassar ebony)

Fig.57:
Light-coloured wood with a cloudy grain (e.g. birch)

Fig.58:
Even, light-coloured wood with a fine grain (e.g. pear, reddish, or maple)

Table 3:
Wood – uses and properties

Tree species	Properties	Use	Processing
Abachi	Soft, light hardwood, low strength, light coloured, straw-yellow, slightly structured surface	Used for making decks in model ships. Veneer strips are used to represent wood surfaces (grooved, structured)	Veneer strips can be easily cut with a utility knife (along the grain). They bond excellently with white glue, or with an extensively applied spray adhesive
Balsa	The lightest of all commercial timbers. Shines slightly, has a whitish surface and a soft, velvety, homogenous structure	Especially suitable for making model aircraft. Veneer strips are used to represent wood; wood profiles to represent buildings	Simple to cut with utility knife or saw. The wood can break easily along the grain; very easy to bond with white glue
Beech	Strong hardwood, fine, uniform grain structure, light brownish-red colour	Suitable for all kinds of models: wooden blocks for representing urban developments; veneers for surfaces and model buildings	With standard wood-processing tools. Can be sawn and sanded. Very easy to bond with white glue.
Linden (lime)	Short-fibred, soft hardwood, plain surface, yellowish, light colour	One of the woods most commonly used in model construction. It is suitable for most purposes	Depending on dimensions, linden is very easy to process with a utility knife and saw
Mahogany	Very hard tropical wood, slightly shiny surface, dark, reddish-brown colour	Used to contrast with all light-coloured woods and modelling materials	Owing to its hardness it can only be processed with a saw and sanding tools

Maple	Soft hardwood, yellowish natural white colour; fine, differentiated grain pattern	Can be used for all types of models, wooden blocks in urban planning models and veneers for surfaces and models of buildings	With standard wood-processing tools, sawing and sanding. Very easy to bond with white glue
Pear	Evenly structured hardwood, light reddish-brown with a fine surface	Suitable for all sorts of models: wooden blocks for representing urban development schemes, veneers for surfaces and model buildings	With standard wood-processing tools, sawing and sanding. Very easy to bond with white glue
Pine	Long-fibred softwood with a typical, striking surface structure, yellowish colour	Its dimensional stability makes it very suitable for wood profiles in frame structures and construction models	Depending on dimensions pine is very easy to cut with both a utility knife and a saw
Walnut	Depending on origins: fine or coarse grain, deep dark brown colour	Used to contrast with all light-coloured woods and modelling materials. Very fine appearance	With standard wood-processing tools, sawing and sanding, very easy to bond with white glue

Wood and tree species

When a modelmaker is working with natural wood, he or she must use a kind that is also suitable for small-scale models. Coarse-grained wood species with distinct markings, such as Macassar ebony and zebrawood, are unsuitable. › see Fig. 56 The surface should look smooth and plain. The woods given in Table 3 are highly recommended.

Wood-based materials

Wood-based materials, which are usually supplied as panels, are made from the waste wood generated by the timber-processing industry. In the building industry, they are used for finishing furniture and interiors as well as for making architectural models. The boards are also used for making bases, such as the loadbearing substrates of models, and for large-scale models of interiors.

Table 4:
Wood-based materials – uses and properties

Wood-based material	**Properties**	**Use**	**Processing**
Chipboard	Reasonably priced wood panel, manufactured from wood shavings and glue, rough surface, material thicknesses: 6.0–22.0 mm	For making mounting boards, for depicting rough and structured surfaces with surface treatment	Processed like wood (sawing and sanding), very easy to bond with white glue. Can warp if exposed to excessive moisture
MDF	Hard, high-density wood fibreboard with homogenous, smooth board structure, high dimensional stability, flat surface, "natural brown" colour or stained in many different colours, including black	For mounting boards, raised-relief models and complete, large-scale (1:50) building models	Processed like wood, very easy to bond, the surface is treated by painting or staining
Plywood (birch, beech, poplar)	Glued-laminated wood panels made of several layers of one of these species, visible grain, same colour as that of the wood used	For raised-relief models and model buildings. Used as an alternative to models made of wood only, as plywood with small cross sections is also stable	Easy to process with utility knife or saw and, in some cases, with a milling machine (birch). Very easy to bond with white glue
Coreboard	Plywood made of glued wood; visible grain, same colour as that of the wood used	Suitable for mounting boards and bases	Processed like wood. Owing to its composition (glued rods) it can generally bear loads in one direction only

Fig.59: Wood as a panel material: birch multiplex boards, chipboard, MDF

Fig.60: Model to a scale of 1:50, made of MDF panels

Fig.61: Model made using MDF (painted black) for the site and linden for the design (contrast in material and colour)

Fig.62: Model made of wood profiles to show the skeleton structure and MDF for the mounting board

METALS

Special care must be taken when representing metals, since their specific properties cannot be faithfully simulated with other materials. If a material is intended to convey the aesthetic idea behind the concept, it will also affect the impression created by the model too. Filigree steel columns and tension ties are ideal for representing thin metal profiles in situations in which the specific properties of other materials would cause these to fail.

Metal sheeting and profiles are used in architectural models. Smooth sheets, in thicknesses of approx. 0.2–4 mm can be used. There are also

Table 5:
Metals – uses and properties

Metal	**Properties**	**Use**	**Processing**
Aluminium	Silvery white, air- and water-resistant owing to thick, opaque, oxide layer that forms on the service, non-magnetic	Aluminium sheets and profiles can be used to represent metal building elements, e.g. corrugated sheet for the roof	Cannot be soldered. Bonded with glues (all-purpose glues), easy to polish, malleable
Brass	Alloy of copper and zinc. Red to light red, depending on the proportion of copper used. A golden colour can be produced by using a higher proportion of zinc	Brass sheets can be used to simulate shining gold surfaces; profiles can be used for loadbearing models (girders and columns)	Can be soldered. Bonds well. Depending on strength of material, it can be processed with metal tools. Polishes well
Copper	The only red metal, oxidizes in contact with the air, turning first red and then green	Copper sheeting and profiles should be used to represent copper in the model	Can be soldered; glues well. Depending on strength of material, it can be processed with metal tools. Polishes well
Iron and steel	Dark silvery colour, corrodes in contact with moisture and oxygen, resulting in formation of reddish-brown rust; magnetic	Steel sheets can be used to simulate metal surfaces, profiles can be used for load-bearing models (girders and columns)	Can be soldered and welded or bonded with all-purpose glues. Materials are corrosion protected (e.g. galvanized) or later painted. Cut with tin snips or metal-cutting saw

Nickel silver	Alloy of copper, nickel and zinc, silver colour and surface, good air-corrosion resistance	Nickel silver sheets are suitable for representing metallic and shining building elements.	Can be soldered; bonds well. Ideal for chipless processing (swaging)
Stainless steel (V2A)	Silvery grey, smooth, fine surface, non-magnetic. The material can be processed to stop it rusting.	Suitable for use in areas exposed to moisture (e.g. exteriors)	Bonded with glues (all-purpose glues)

structured sheets such as channelled plates, corrugated sheeting and chequered plates; and sheets with round-hole, square-hole or elongated perforations, as well as expanded meshes.

To represent steel profiles realistically, it is also possible to buy miniature solid round profiles, round tubes, T-, L-, H- and I-profiles, square and rectangular profiles.

Fig.63:
Model of a building with a façade of very fine perforated sheeting

PLASTICS

There are so many different plastic products that it is difficult to describe them in general terms. Most plastics are malleable synthetic materials made of macromolecules. Carbon (an organic material) is one of the main constituents. All plastics can be processed easily and with a high degree of precision. Their light weight and stability combine to make them very useful in areas where other materials would be out of the question.

Polystyrene models

The most common plastic used in modelling is polystyrene (PS). Being mass-produced, polystyrene – like polypropylene (PP) and polyvinylchloride (PVC) – is inexpensive and has a wide variety of applications. Many architects and modellers work with nothing but polystyrene. As a result, a special type of design and presentation has been developed for constructing architectural models. Polystyrene is white and smooth. It is ideal for doing precision work and making filigree forms. Models made solely of polystyrene offer the desired degree of abstraction and also provide plain and simple examples of three-dimensional architecture.

Nearly perfect results can be obtained using plastics, a feature that sets them apart from other materials. It is possible to work precisely to within fractions of a millimetre, which is a great advantage when making urban planning models, for example. Furthermore, it is possible to

Fig.64:
Plastics – a wide range of choices for modelling

simulate transparent elements such as glass by using thin, transparent PVC foils. Plastics are an indispensable modelling material. Polystyrene is the material most widely used by architects in competition projects and in modelling in general.

Acrylic glass in modelling

After polystyrene, acrylic glass (chemical name: polymethyl methacrylate, PMMA; also known as perspex and plexiglass) is the plastic most widely used in architectural modelling. Being a thermoplastic material, it displays excellent thermal ductility. Furthermore, many of the varieties manufactured are ideal for representing glass panes, as well as glass and transparent building elements (e.g. in urban planning models). The surface can be modified. Satin matte structures can be made by grinding acrylic glass (so-called "wet grinding", with fine-grain sand paper, for example grain 600). Grid structures and patterns can be made in acrylic glass by milling, grooving or scoring it with a sharp utility knife blade.

There is a very wide range of plastics and plastic products. In modelling, too, far more types are used than are described here. The only criteria for using plastic are the form and colour desired and its surface properties. Among the plastic products available for modelling in the form of panels, foils and profiles, special mention should be made of polyester and polyethylene (PE plastics).

In addition to the above-mentioned "raw materials" for making architectural models, a number of other products – paints and varnishes, for instance – are also indispensable modelling materials.

\\Hint:
In modelling, PS rigid foam is often finished in a different colour. Unfortunately, some paint solvents – especially if sprayed – react with the plastic compound and can even dissolve it. It is therefore advisable to carry out tests first.

\\Tip:
Acrylic glass is often very expensive, and difficult to obtain from specialist dealers. It is therefore worth contacting processing companies or the manufacturer directly and asking for waste materials or samples.

Table 6:
Plastics – uses and properties

Plastic	Properties	Use	Processing
Polystyrene (PS): as a duroplastic, "hard" plastic	Shock-resistant, hard plastic. Matte, white, opaque panels are used in modelling. PS is not UV-resistant. Material thicknesses 0.3–5.0 mm	Universal: in all areas of modelling	Very easy to cut with a utility knife, the surfaces are easy to polish and very easy to mill. They can be easily bonded together using solvents or special PS adhesives and contact glues. Easy to paint or varnish
Polystyrene hard foam (e.g. Styrofoam)	A porous material manufactured in panels and blocks. Not shock-resistant. The surface can be easily depressed. Available in different colours, depending on the manufacturer.	For building and urban planning models	Very easy to cut with a hot wire cutter, and to cut or carve with a utility knife. It can also be polished and painted
Polypropylene (PP)	Heatproof, hard and tear-resistant plastic. In modelling, it is generally used in the form of thin, transparent or opaque foils. It has a non-scratchable surface, is UV stable, and available in various thicknesses	As a translucent foil, it is excellent for imitating matte glass surfaces and for designing lamps	Very easy to cut with a utility knife. It can be bent, folded, grooved, welted, moulded and punched. Before bonding, it must be pretreated, e.g. with Poly-Primer

Polyvinyl-chloride (PVC)	Depending on the mode of manu-facture, either transparent or opaque, various material thick-nesses	Transparent foils make good imitation glass in models, and can be used as thin foil in many different areas	Very easy to cut with a utility knife. It can be drilled, milled or turned. PVC surfaces can be bonded together with standard plastic glues and contact glues
Polycarbonate (PC)	Very strong, impact-resistant plastic, weather resistant, fine surface, transparent or milky-translucent plastic foils	Transparent foils make good imitation glass in models, and can be used as thin foil in many different contexts. It generally has a finer and smoother surface than PVC	Very easy to cut with a utility knife. Thick sheets can be scratched and broken. PC surfaces can be bonded together with solvents and contact glue and leave no residues
Acrylic glass (PMMA)	Very transparent and bright, excellent optical properties, similar to glass. Weather-resistant plastic, available in transparent, translucent or opaque form	As a trans-parent material for repre-senting glass or water	Thin foils can be easily cut with a utility knife, thicker materials must be broken or sawn; moisten before grinding and polishing. Can be easily bonded with solvents, contact glues or special glues for acrylic glass

Fig.65:
Monochrome polystyrene model, sprayed white with latex paint

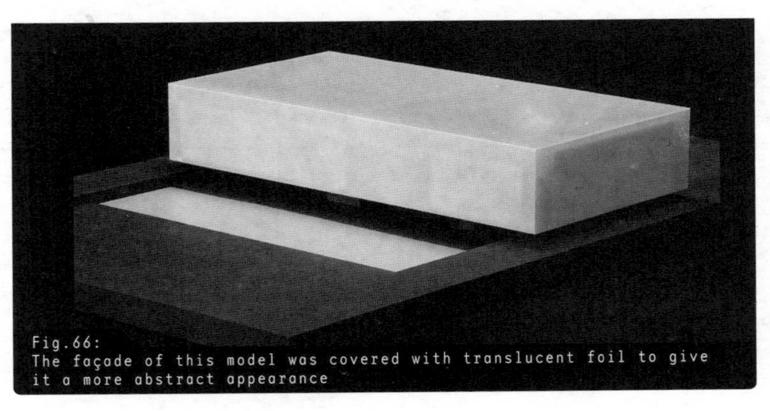

Fig.66:
The façade of this model was covered with translucent foil to give it a more abstract appearance

Fig.67:
Urban planning model with buildings made of rigid expanded polystyrene

PAINTS AND VARNISHES

The surface of a material is not always the surface that will be visible on the finished model. Coats, films and paint finishes may be applied to many different grounds in order to modify the original design. These products also improve the resistance of the surface (insolation, UV light resistance, water-resistant surfaces, preventing yellowing).

Table 7:
Paints and varnishes – areas of application

Paints and varnishes	Use	Result
Paint sprays (colouring pigments with adhesives and propellents)	Can be used with most materials. However, care must be taken that chemical reactions do not take place between the solvent and the material (e.g. compatibility with polystyrene foam boards)	Different types of sprays can be used (silk matte or gloss) to match the colours of the surfaces
Clear varnishes	Can be used with many different materials. Applying clear, transparent layers, you can create nuances without the need for additional materials	For example, nuances in sites, topography and the building, varying intensities of colour or degrees of brightness on the surface of the material. Varnishes conserve the surface (e.g. of Finnboard, preventing the material from yellowing)
Acrylic paint	For painting specific building elements or surfaces on the model. It is easy to mix and the brush can be cleaned with water	The different materials can be made to match by applying a uniform monochrome coat; very intense, bright colours (depending on the pigmentation)

Wood stain	Stains can be used to change the natural colours of the wood surface. The original properties, such as the grain, are preserved	By changing the hue, different shades and nuances can be created. Coloured stains make the material look very different.
Oils and waxes	The surfaces of both wood and cardboard can be treated with these products, thus changing colouring and brightness	These protective surface coats can often be applied to intensify the impact of the surface properties of woods

PLASTER, CLAY AND MODELLING CLAYS

Materials such as plaster (gypsum) and clay, like the other materials presented here, form a distinct category of materials. People often limit themselves to using only the material from which the entire model is made. These materials are particularly suitable for presentations that focus not on the exact formal details but on a building's corporeality and massivity. Plaster, it might be added, can be cast in very precise, smooth forms.

Plaster

Gypsum is used as a building material on a scale of 1:1 (in the form of plaster and stucco, for example). Its chemical name is calcium sulphate, a compound that occurs naturally and as by-product of power stations.

In modelling, it is used as a pouring compound. First, a mould must be made. This is a time-consuming process, but its advantage is that you can cast a product as often as you wish.

\\Tip:
When you use paints, varnishes or other surface coats, it is worth doing tests first to see what effect they have. Using samples and testing materials, you will acquire a feel for the products' properties and for different ways of using them. Experimenting can be good fun too. Colours and contrasts are best judged under natural lighting.

Fig.68:
An urban design model made of fired clay

Gypsum, as "modelling plaster", is available as a white powder. It is mixed with water and used either in liquid form or applied in a thicker consistency with a putty knife.

Liquid plaster quickly sets and hardens. Once it is in this state, it can be sandpapered and painted or varnished.

Clay

Clay has long been used to manufacture a wide variety of objects in different fields.

The history of architectural models shows that clay (also known as loam) has been used in certain areas since models were first built:

- For the construction of forms: free forms, organically formed objects
- Sculptures
- Three-dimensional solid models, e.g. for urban development projects

The great advantage of clay lies in the fact that it is simple to use. Working with clay is also a tactile experience. With clay you can create "corporeal" three-dimensional, sketches, work on entire development processes and make changes quickly and effortlessly. While you are processing clay, make sure that you always keep it sufficiently moist (with water) to prevent it from drying out and cracking.

When a clay model is ready, it has to be fired in kiln to preserve its form. The product is known as pottery or ceramics. It is used the building industry for making bricks and clay roof tiles.

Plastic modelling clay

In addition to the classic materials plaster and clay, other products are also available:

- Air-drying modelling clay: its processing properties are similar to those of clay, with the difference that it hardens in the air and can be subsequently processed with sanding tools – like wood.
- Plastic modelling clay/ plasticine: a modelling material that basically remains processible and formable, whilst preserving a constant degree of hardness. A typical feature of this material is that it becomes "softer" at higher temperatures, but retains a relatively firm consistency at normal room temperature. Plasticine has a familiar greyish-green colour. Plastic modelling clay is available in many different colours.

Assembly units are increasingly often used in modelling.

ACCESSORIES: TREES, FIGURES AND CARS

When clients first set foot in a new building, they often exclaim: "I imagined it would be much smaller (or bigger)." This is hardly surprising, since the only impressions they have of a project's scale before it is built are gained from drawings and models. And this is the crux of the matter: how can architects convey an accurate impression of the size and scale of the final structure? To give the viewer an idea of the real size, they can use "known quantities". Such reference points give the viewer a basis upon which he or she can imagine the finished product. Among the objects that convey an idea of scale are small-scale human figures. We all have an idea of our own height and can relate this to the model. If we read and grasp objects in relation to this "human scale", we find them easier to judge.

Accessories

In addition to figures, there are many other objects that help us visualize the scale of a project. Vehicles, available from specialist dealers, or represented by other objects, are indispensable in a multi-storey car park. The same applies to trees and plants, which must be shown if they are an

Fig.69:
Materials used to represent trees

important feature of the concept (in landscape planning or urban design projects). Balls, rods and simple, three-dimensional blocks can be used to represent trees. However, the best method is simply to use trees themselves. Various dried plants (such as yarrow), which are normally used in flower arrangements, make ideal miniature trees. If trees play an important role in the model design (site design, relationship of buildings to trees and the interior courtyard etc.), the abstract trees must be selected very carefully, since they are often decisive for the overall impression.

In architectural models, the accessories mentioned above are normally all you need. In some cases, however, other objects may be more suitable for illustrating the purpose of the future building:

_ Streetlamps as a line-forming element
_ Street furniture (benches, advertising pillars, signs)
_ Furnishings in interiors such as shops and restaurants
_ Trains, ships and aeroplanes, in the case of transport buildings

\\Example:
Modelling figures are available in all scales as injection-moulded products. For the common scale of 1:200, grains of rice can be stuck vertically into the base as an abstract representation of people.

Fig.70:
Wooden balls

Fig.71:
Branches/twigs

Fig.72:
Foam

Fig.73:
Foam

Fig.74:
High-resistance foam

Fig.75:
Brush trees

Summary: Most of the materials mentioned above are obviously not specifically manufactured for making models. It is up to you, as the modeller, to use your imagination and apply unusual materials to the context of architectural models.

\\Hint:
When you choose suitable materials, consider how long you want to use or keep the model. Many materials age – in this respect models are very similar to constructed buildings – changing the appearance of models as time goes by. Sunlight (UV radiation), in particular, darkens cardboard and wood. Finnboard, for instance, turns yellow very quickly. Although ageing affects the aesthetic appearance of wooden models, it has a far more detrimental impact on the durability of cardboard models.

FROM DRAWING TO MODEL – STEPS AND APPROACHES

Below, we shall examine different methods of making models and show you different ways in which you can make one yourself. These tips are intended as a guide to help you in your creative work.

A FEW PRELIMINARY THOUGHTS

Objectives of the presentation

Before you pick up your utility knife and cutting ruler, you should consider your objectives. Although modelmaking can be a very creative way of designing a building and a very pleasurable activity (e.g. tearing and joining pieces of paper), if you want to make a model for representational purposes, you should avoid proceeding by trial and error from the very start. Hence, you should begin by clarifying your goals.

You should first decide what you want to present. If it is a design principle (e.g. a stone slab with a glass hall), then this will influence your choice of scale, materials and the level of abstraction. If you want to present a concept to, for example, a professorial committee or representatives of building clients, you must decide exactly what you want to communicate to the target group. You should also consider whether they are likely to be very imaginative. In general, people teaching architecture have far more experience in judging designs, and are more likely to be familiar with abstract models than, say, building clients who have had little experience in the field of construction.

Relationship between plans and model

Furthermore, the presentation plans will influence the choice of material and the format of the mounting board, especially if it and the model are meant to form a coherent whole. Hence, you should carefully select the essential elements and graphically explore different levels of abstraction – using sketches, too, if necessary.

Deciding plan detail

The degree of detail shown in your model will depend on spatial relationships and the physical size of the mounting board you intend to use. If some parts of the design (e.g. a striking axis in the immediate surroundings) are very important, but too big to fit on a suitable base, you might have to reconsider the scale of your model.

THE MOUNTING BOARD

If the surroundings and the terrain are to play a decisive role in your concept, you should start by thinking about the mounting board. Although this may not be appropriate for every project, it is always a good place to start. Begin by making a working model of the location. This can, of course, be used for the presentation too. Often, when people start working on a design, they are short of ideas, so it is often worth reproducing the terrain and topography and making a spatial analysis of the site. Very often, the best ideas arise when you get down to work. Another reason for reproducing the terrain at the beginning is that the workload generally tends to grow toward the end of a project. It is therefore worth preparing and finishing this part of the work before you do anything else.

When making the terrain (the "foundation") make sure that the dimensions of the whole model are right: if the model is too large it will be heavy and difficult to transport; if it is too small, it will not communicate enough information during the presentation. Boards manufactured from wood-based materials usually make good foundations. › see chapter Materials

If you already have a suitably sized mounting board for presenting the section you have chosen, you will now need a site plan to transfer the surroundings or the terrain to the mounting board. Once you have done this, you will have to transfer your geometric planning data onto the modelling material. You can use one of the following methods:

_ Frottage: copy the drawing as a mirror image, affix it to the material with adhesive tape and then rub it onto the material with the aid of acetone solvent. In the process, the printing ink is transferred from the drawing to the material.
_ Tracing with needles: mount the drawing on the material, pierce the key features into the material with a needle or a pounce wheel (a special modelling tool) and then trace the line geometry.
_ Drawing: the best way of making simple outlines is to draw a line on the material with a fine pencil.

\\Tip:
Transferring the typography of the terrain to the model is generally very time-consuming. If you are making a raised-relief model, select a material thickness that, in the desired scale, corresponds to the elevation lines on the terrain plan. This approach renders further changes unnecessary. First, make the required number of layers to fit the dimensions of the mounting board and glue an entire layer on to the mounting board. Now cut the first elevation layer out of the site-plan base, which has also been cut to size, and transfer it to the next layer of paperboard. Once you have glued the first layer, cut additional layers and transfer them one at a time to the others until you have created a perfect reproduction of the terrain. Note that you may have to create a new model of the terrain of the new construction site, which will then have to be integrated into the raised-relief model.

MAKING INDIVIDUAL BUILDING ELEMENTS

Modelling is more like constructing a prefabricated building than erecting a complete structure (from the primary construction to the finished product). Generally, it is advisable to finish building elements completely before you join them to others. Typical elements include:

_ Floor plate, ceilings, roof top and all of the structure's horizontal building elements
_ Walls, shear walls and façades, including the arranged window and door apertures as well as all vertical elements
_ Interior walls

Working from plans and sketches

These elements are made on the basis of architectural drawings (in the same way that a building is constructed on the basis of a building plan or work drawing). All we have on paper are the classic two-dimensional projections, e.g. the plan, the section and the elevation. However, the plans and the sections are equally important. They each provide two-dimensional views of a building. Together, however, they create a three-dimensional impression. The best method is to transfer the geometry of the plan to scale onto the material. There are different ways of doing this:

_ The walls and supports are mounted as indicated on the plan (three dimensions). This is a quick and easy method, regardless of whether you are using working models or urban design models.
_ All the important information in the plan is transferred to the desired material. › see above

Once the information in the plan has been evaluated and applied layer by layer, the height indications in the longitudinal sections and cross sections gradually assume concrete form.

\\Tip:
When you transfer the lengths and heights of the elements to the model, pay attention to the thicknesses of the materials. If you want to mitre a wall corner, both sides must be made to their full length. If the elements are to be butt jointed, however, you must deduct the width of the material of the projecting element.

Fig.76:
"Stair boards" cut to the right width

Walls are usually the most important parts of a building. When you are constructing models, you should design them not as single elements, but as part of a whole, and make them accordingly. In this respect, modelmakers can work far more efficiently than builders, since they can cut or saw walls into endless strips and then cut them to the required length when needed.

This method of "endless manufacturing" greatly simplifies the modelmaking process for both urban construction projects (involving rows of houses or buildings with the same cross section as "endless rows"), and for other building elements with the same cross sections (supports, window frames, floors etc.). It is worth making a large number of individual elements and then varying them, especially if you are experimenting with your model.

Stairs

Stairs are often a decisive element in any project and their representation is very important. In working models, stairs are simplified and installed in the form of slanted surfaces (ramps). This work can be done quickly, and everyone knows what the ensuing slopes are supposed to

\\Example:
When you are making stairs, consider how many staircases you are going to have in your model. You can save yourself a lot of work by making the flights (in the case of single flight stairs) much wider and cutting off single sections in the desired width with a utility knife or a saw.

Fig.77:
Transferring the drawing to paperboard

Fig.78:
Cutting out the mounting board

Fig.79:
Measuring out the walls in strips

Fig.80:
Cutting out the apertures

Fig.81:
Gluing the walls together

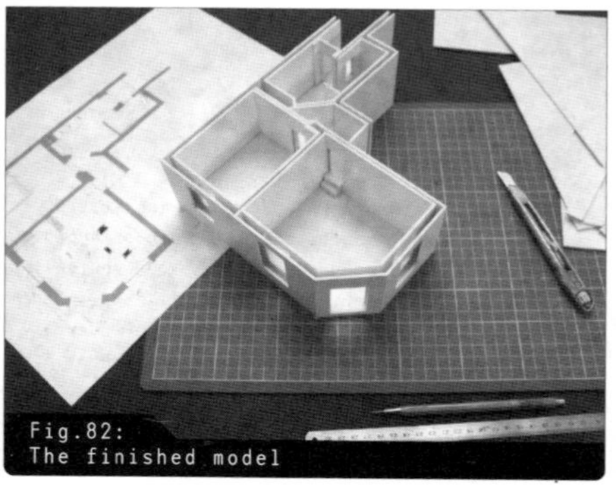
Fig.82:
The finished model

represent. Representing a real staircase with treads and risers is more time-consuming, but it also creates a more impressive result.

ASSEMBLING THE ELEMENTS

The elements should also be assembled systematically since in most cases you cannot take them apart without causing damage once they have been glued. An effective method is first to mount two exterior walls of a building and then successively glue the interior walls and the ceilings. Finally, you can complete the building by attaching the missing exterior walls.

A detailed simulation of windows and façade elements is also important for the expressive power of a building model. You should install the windows before assembling the exterior walls and façade. The same holds true for ceilings with openings for stairs and for walls and roofs with their apertures.

At times it may make sense to prefabricate all the elements associated with a specific phase of construction – i.e. entire building sections – and to assemble the entire unit afterwards. However, you should first make sure that, should the need arise, you can adjust the connected elements in case the building sections do not fit together with geometric precision. ›✎

FINAL TASKS AND ACCESSORIES

After all the elements have been assembled to form the building, a few final tasks remain before the model is complete. The process must be carefully planned, especially if you want to paint individual elements or

✎

\\Tip:
You should avoid using a utility knife on a model that has already been glued together since you can never work as precisely as you can on a cutting board. A better approach is to prefabricate as many elements as possible and then glue them together. You can check for measurement errors by temporarily joining the parts together. You can ensure that your model is stable by setting up the pieces at right angles.

the entire model. If, for example, a piece running through the entire building is to be a certain colour, you should make this piece and glue it together as a single unit so that you can paint it independently of the rest of the model, using a spray gun if necessary. If parts of the model are a certain colour, you should also consider whether it is best to add accessories such as trees and figures before or after painting.

Adding accessories

It is definitely worth considering whether, and how, to use accessories as the model nears completion. Trees are a very popular element and can substantially improve an architectural model. Even so, trees and other accessories should only be included if they are relevant to your concept, but not if they merely serve as ornamentation. Often, modellers make the mistake of "planting" too many trees on the mounting board, which can divert attention from the actual subject of the model. As a general rule, if trees are architectural elements – that is, if they have a space-forming function – they should be part of the presentation, but they are not absolutely necessary (less is often more).

›✎

PRESENTATION

The finished model will usually be part of an exhibition or presented in conjunction with a talk. You should not leave this last step

✎

\\Tip:
If the accessories play too prominent a role, they may detract from essential features of the model: viewers will literally not see the architecture for the trees. To minimize this effect, you can paint the chosen materials to match the other materials used. Not only will the elements blend in better with the surroundings, but they will still be perceived in a spatial context. If you execute large public buildings such as concert halls, museums and the like as small-scale models, you can add small abstract figures to indicate paths, emphasize public spaces and give viewers a better understanding of the real context. If you want to create the impression that the structure exerts a magnetic pull, you might consider adding more concentrated groups of figures near the main entrance than those scattered across the grounds.

to chance. When pictures are hung on the wall, they are positioned at eye level to achieve the best possible effect. Models should also be presented in such a way that audiences can view them from the right angle. Whereas urban design models are usually viewed from above, you should ensure that people viewing an interior model can look into the spaces without assuming contorted or uncomfortable positions.

Constructing a base

To ensure that the model is exhibited at the right height, you can build a base (e.g. in the form of a pedestal) that conforms to model's shape. A base can be used to support a model conceptually: for instance, if a tall, narrow building is displayed, a tall, narrow base can illustrate the architectural concept from a distance.

The base of a group model is often equipped with wheels so that it can be moved easily from presentation to presentation. All you need to do at the new venue is replace the inserts.

Display cases

High-quality models can be presented in a display case to protect their surfaces and emphasize the object-like character of the architectural representation. A perspex case protects the model from damage at exhibitions and is particularly useful at public shows.

Hanging models on the wall

Urban design models (raised-relief models of landscapes) can be hung on the wall like a drawing or a plan (a kind of three-dimensional plan). Here, it is important to find out in advance what kind of wall is available. If plans are hung on partition screens, they may not be inordinately heavy.

›✎

✎

\\Tip:
You will probably be reluctant to throw your models away, but in the course of your studies, you will make so many that you will probably not have the necessary space to store them all. Since models stored in basement or attic spaces can be damaged by moisture and fluctuating temperatures, you might consider hanging your models on the wall after presentation. In this way, you can both store them and use them as wall decoration.

IN CONCLUSION

The study of architecture confronts students with a plethora of demands and challenges. Among other things, they must learn to make good models. Many of these models are created for presentational purposes just before an assignment is due. The potential they offer as design and work tools is overlooked. Studying architecture involves learning about technical matters as well as artistic and creative processes. Models are a way to explore space and proportions and to promote three-dimensional thinking. Models show the ramifications of paper sketches and also help students train their ability to imagine the spatial relations of two-dimensional drawings.

Models are also a method of communicating with a lay audience. If a client hires an architect to design a building, it is ultimately the model and not the detailed plans that will give him or her a concrete idea of the architect's "product". The model provides a foundation for approving and rejecting ideas, making improvements and fleshing out details. The different roles that the level of detail plays for architects and clients are reflected in the ways they respond to the model: clients see the model as a finished result and want a high level of detail because they are eager to see the project completed. Architects, on the other hand, prefer a model that is only moderately accurate since it will allow them greater creative freedom.

For both, models round off the ways a project can be represented. As the three-dimensional implementation of a design idea, a model is every bit as important as presentation drawings.

APPENDIX

ACKNOWLEDGEMENTS

Special thanks go to:

_ Chair of Building Theory and Design, Professor Arno Lederer/ Professor Daniele Marques, University of Karlsruhe (Photographs of models that were built over the last few years for design and end-of-term projects in design theory. Master model-maker Manfred Neubig provided modeling courses to support the design process.)
_ Gerstäcker-Bauwerk GmbH, modelmaking materials and art supplies, Thomas Rüde and Marc Schlegel, Karlsruhe (provided advice on materials and assistance when tools and modelmaking materials were photographed)
_ Wood workshop at the Department of Architecture, workshop supervisor Wolfgang Steinhilper, University of Karlsruhe (assistance with photographs)
_ Metals workshop at the Department of Architecture, workshop supervisor Andreas Heil, University of Karlsruhe (assistance with photographs)
_ Christoph Baumann, Karlsruhe (model photographs)
_ Verena Horn, Karlsruhe (assistance with photographs)
_ Peter Krebs, Büro für Architektur, Karlsruhe (model photographs)
_ Stefanie Schmitt, Stutensee (model photographs)

PICTURE CREDITS

The following photographs were generously provided by the Chair of Building Theory and Design, Professor Arno Lederer/Professor Daniele Marques, at the University of Karlsruhe. Professor Lederer supervised the designs. The research associates were Kristin Barbey, Roland Kötz, Peter Krebs and Birgit Mehlhorn. Manfred Neubig oversaw the modelmaking that accompanied the design process. If not otherwise indicated, photographs were taken by Thilo Mechau, on staff at the photography workshop of the Department of Architecture.

Fig. 1 Student project, design, Reykjavik harbour, Florian Bäumler

Fig. 5 Student project, design, Palace of the Republic, Ruwen Rimpern
Model of the urban surroundings: Manfred Neubig

Fig. 6 Student project, design, Reykjavik harbour, collaborative urban design model

Fig. 21 1st row: student project, design, "Do it again," Steffen Wurzbacher
3rd row: student project, design, "Do it again," Matthias Rehberg
4th row: student project, design, "Do it again," Ioana Thalassinon

Fig. 24 End-of-term project for building theory course, collaborative model made of plasticine, photo: Cornelius Boy

Fig. 26 Left: student project, design, Palace of the Republic, Marc Nuding
Right: student project, design, Islamic community centre, Axel Baudendistel

Fig. 27 Master's project, seaside resort in Barcelona, Philip Loeper

Fig. 48 Student project, design, Reykjavik harbour, Holger Rittgerott

Fig. 49 Student project, design, "Do it again," Lisa Yamaguchi

Fig. 50 Student project, design, "Do it again," Andrea Jörder

Fig. 51 Student project, design, "Do it again," Lisa Yamaguchi

Fig. 60 Student project, design, "Do it again," Benjamin Fuhrmann

Fig. 61 Student project, design, Reykjavik harbour, Markus Schwarzbach

Fig. 62 Student project, design, Reykjavik harbour, Kuno Becker, Tina Puffert, Holger Rittgerott

Fig. 66 Student project, design, Palace of the Republic, Matthias Moll

Fig. 67 Student project, design, Reykjavik harbour, collaborative model

Fig. 68 Student project, design, Islamic community centre, Axel Baudendistel

The following photographs were made available by Büro für Architektur Peter Krebs in Karlsruhe.

Fig. 4 Competition model, Mainhardt community centre

Figs 12, 13, 43 Competition model, Church of St Augustine, Heilbronn

Fig. 14 Competition model, Stella Maris Chapel, Stuttgart

Fig. 25 Competition model, St George's parish hall, Riedlingen

Fig. 65 Competition model, community centre, Schwetzingen

The following photographs were made available by Christoph Baumann, Karlsruhe:

Figs 70, 71, 72, 73, 75 Details from study models

The following photographs were made available by Bert Bielefeld, Dortmund:

Figs 11, 16, 45–57, 74

The following photographs were made available by Isabella Skiba, Dortmund:

Figs 8–10, 44

The following photographs were made available by Stefanie Schmitt, Stutensee:

Fig. 18 Study model, St Francis Kindergarten

Fig. 19 2nd row: Master's project, Olympia station, Stuttgart

Fig. 63 Master's project, Olympia station, Stuttgart

The author has supplied Figs 2, 15, 17, 20–23, 28–42, 52–59, 64, 69, 76–82

P10

作为一种表现方法的建筑模型

什么是模型制作?

在意大利文艺复兴早期，模型制作发展成为最重要的建筑表现方法。它不仅仅是建筑制图的补充，而且还经常是表达理念与描述空间的最主要方法。从那时开始一直到现在，建筑师、工程师与委托人都开始使用建筑模型来表现建筑设计。

平面图（设计草图，例如技术平面图）与建筑模型都是用来描述建筑与空间的手段，然而图纸只能够表达两个维度。一旦构思或者方案被设计出来并且绘制成草图与细部图，就应该开始在空间上进行处理。虽然说只有建成的建筑才能够表达出完整三维效果的感受，模型却可以预见随即产生的建造过程。这样看来，模型在缩小的尺度上再现了建筑。用模型来进行工作是十分重要的，尤其是在学习建筑学的过程中，因为学生通常很难有机会将自己的设计建造起来。

动机：为什么制作模型?

一个模型不必完整与顺利地实现一个设计或一项建筑任务，但是它却仍能在许多方面成为有用的工具：模型的缩小尺度，使得它可以用以检验设计并且使设计者获得空间、美学与材料的感受。模型的其他优势还包括其交流与说服的能力：模型可以帮助设计者向自己与他人展示其理念与设计的品质。另外，模型还可以在建造之前用作评估建筑的控制机制。

工作模型——三维草图

建筑学的学生在他刚进入大学的第一年中经常会面对模型制作的挑战。他们会很快地将建筑模型看作是他们儿时制作的模型——火车、飞机和轮船——这些与他们课程学习的要求是完全没有关系的。火车布景中的绿色草坪对于教授与讲师来说是没有任何意义的。孩提时期的玩具现在已经成为职业的现实。

那么学生们该怎样来做？让我们来假设已经给他们布置了设计任务并要求制作模型，预先给定了比例，这会对他们计划采取的解决方案产生很大的影响。在他们的设计工作中是否运用模型在很大程度上取决于他们怎样进行设计，同时也取决于设计怎样进展。如果空间结构的复杂程度远远超过两维的图纸所能表现的范围，模型就成为可以描述它的惟一方法。简化模型，也就是所谓的工作模型，可以帮助设计师找到解决问题的方法并检验设计的理念。那些不能够在模型上实

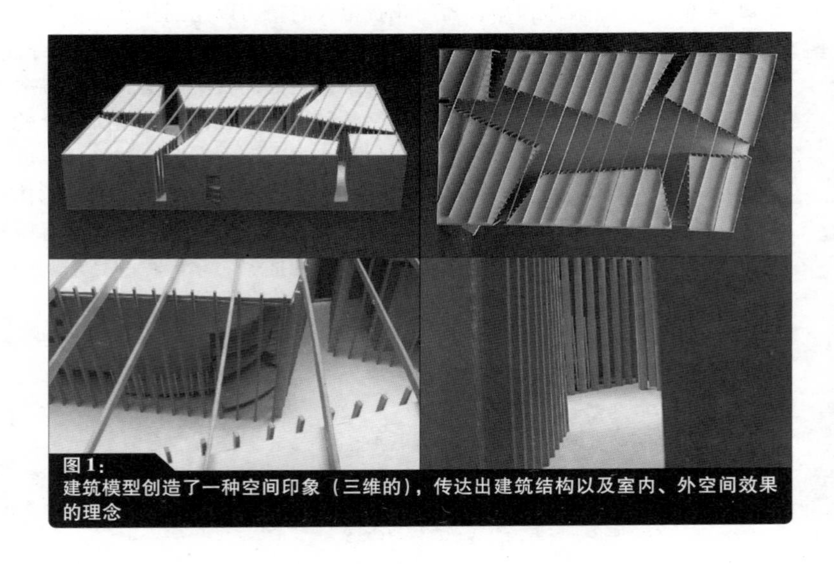

图 1：
建筑模型创造了一种空间印象（三维的），传达出建筑结构以及室内、外空间效果的理念

现的理念就可以放弃了。

工作模型伴随着整个的设计过程。一旦这个过程完成了，下一个目标则变成表达设计中最本质的想法与概念。必须使观赏者相信无论是在概念上，还是在计划的解决方法上这个方案都是可行的。

提示：

工作模型之所以获得这个名字是因为设计师可以与它一起“工作”。它们可以，也应该可以被修改，即便这样做会损害其质量。因此，模型的制作最好可以容易拆分并重新组装。学生们可以使用大头针或可以轻易除去的橡胶粘结剂来将各部分组装在一起。而一旦两部分被粘结在一起的话，它们就被赋予了特定的形式，这就已经不再是工作模型的目的了。

小贴士：

工作模型，可以支持试验性的设计向前推进，而且如果它们足够精致的话，也可以用作表现。例如建筑构件或地形叠层等元素应当临时装置，这样既可以确保模型可以被修改，也可以避免工作中不必要的污染与损坏。而在这一过程后期的着色工作则是丰富工作模型的一种很好的方法。

表现模型——描述与说服

表现模型，需要花费很大的精力以达到几乎完美的程度，它标志着设计过程的完成。在大学中，这种类型的模型用于表现设计想法——概念。在建筑竞赛中，它描述着所提出的解决方案并与其他的参赛方案进行竞争。在这两种情形中，模型都是所提交的建筑方案的补充。也会经常使用新的媒介来表现方案图：计算机生成的三维图像可以为将来结构的样子提供非常真实的渲染图。照片的仿真表现是另外一种模拟空间试验的现代方法。一个模型不可能具有所有的这些功能。它总是自己所描绘的现实的抽象。它惟一的真正功能即是将勾画出来的想法转化为三维的形式。

小贴士：

当有提交设计方案的压力时，完成表现模型总是很困难的。通常更有效的方法是进一步地制作局部模型，或者将基地模型既用于工作模型也用于表现模型。一个表现模型不需要完美，也不需要用最好的材料来制造。而对于一个有说服力的模型来说，所有的这些都是需要的。

P13

模型的分类

抽象——小型化的诀窍

一个模型或多或少是抽象的，在缩小的尺度上表现着现实。但是在模型的制作过程中抽象意味着什么？

抽象的反面是具体。在绘画中，“具体”指的是尽可能精确地描述一个物体。相反，应用于建筑模型中的抽象是要将人们关注的焦点转移至主题，转移至物体所描述的信息的价值以及空间构架上。成败的关键不是对现实的精准描述，而是简化的过程，是将人们的视线导向模型的本质特征的过程。至关重要的是找到恰当的抽象形式，这种形式会反映所选定的比例。这一点可以通过窗户的例子来加以说明：在1∶200的比例上，一扇窗户通常被表现成在选定材料的表面上精确切割出的孔洞。而在1∶50的比例上，窗户则可以更加明显地被观察到。玻璃通过透明的材料加以再现，而窗框则由小木条制成。另一个例子是立面的覆材：在非常小的比例上，它根本不可能被表现出来，而在较大的模型上这一因素的影响会增大，并且也包含在设计内容的范围内。通过选择恰当的材料可以非常真实地模仿立面。

比例与表现之间的关系

在一个方案的开始，模型的制作者就必须选择其模型的比例。建筑物在什么尺度上可以被表达出来决定了比例在其中所扮演的角色。依据比例与抽象的程度，存在着许多不同种类的模型，将在下面的文字中进行解释。

例子：

尝试在缩略模型上再现表面的特征与感觉。粗糙的或木纹的表面可以通过砂纸打磨或"键控雕刻"（keying）卡纸板与木材来模仿。

构造可以通过以下方法来仿造：

— 木制覆层可以通过使用小片木材模仿真实设计的方法表现；

— 砖立面可以通过在材料表面切割出接缝的方法来模拟；

— 支撑构件与梁可以做得与真实形状相同。

P13

概念模型（没有具体的比例）

空间形式通常可以用概念模型加以描述。在这里，设计背后的想法或创造性的概念都可以以完全抽象的方法三维地表达出来（例如将象征作为基本手法）。材料、形式与色彩可以用来强调结构与创造构图。例如，在设计过程的开始，模型就可以将城市空间分析的结果变得可视化。通过在空间并且是在抽象的层面上探究一个课题或一个场所，建筑师可以改变或改善该场所的景观。而模型则可以支持这种方法。

P14

城市设计与景观模型，基地与地形
（比例 1：1000，1：500）

这一种类的模型表现的是城市与自然环境。它是表现过程的第一步，因为它表现了与现有环境之间的关系。在城市空间中，当加入一

图 2：
不同比例下的窗户

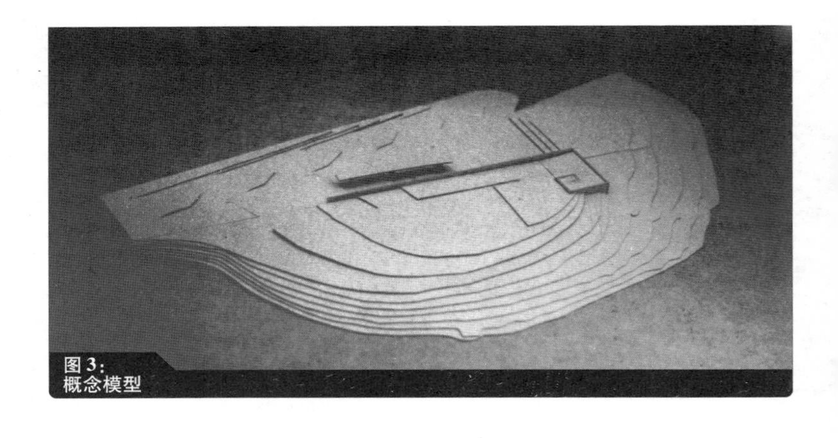
图 3：
概念模型

个新建筑时，表现环境发生了怎样的变化是非常重要的。

这一类模型的特征是高度的抽象化。建筑被简化为“建筑体块”——为了抽象地表达结构，它以高度简化的方法表现建筑形式与三维物体。即便如此，模型还是包含着建筑物的典型特征，例如退进、突出、凸窗与屋顶设计。在这种抽象的形式中，基地——缩小比例的景观——在同一个水平面上，用选定的材料简化地描述。如果地形是倾斜的，在模型中则是采用将水平层面相互堆叠起来的方法。

浮雕模型

如果模型试图表达不平坦的地貌，制作的第一步就是将不规则的自然地形作为水平层的堆积来考虑。材料分层越精细，结果的模型就越精确与逼真（每层的厚度：1.0mm 或 2.0mm）。这项工作是以表现

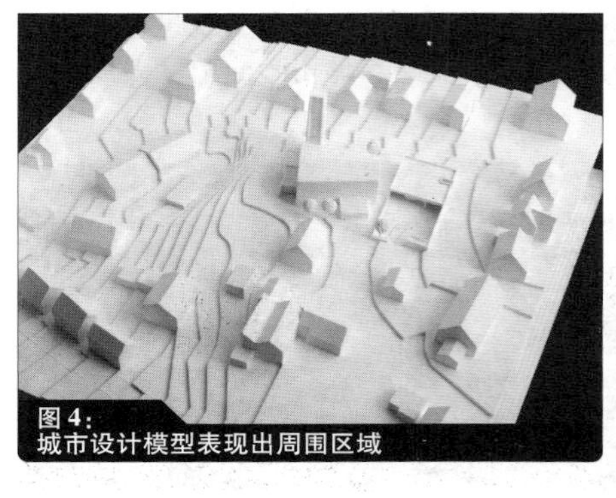
图 4：
城市设计模型表现出周围区域

图 5：
柏林中部（Berlin-Mitte）城市设计模型

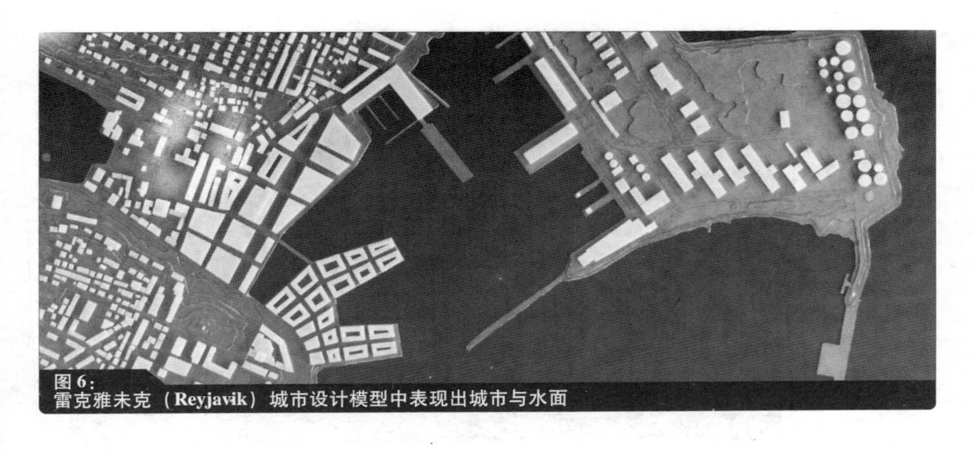

图 6:
雷克雅未克（Reyjavik）城市设计模型中表现出城市与水面

等高线或者至少是能够提供标高信息的平面图为基础的。一旦掌握了真实的地形区位，就可以绘制出等高线（曲线、直线、折线）。在将各层相互叠加在一起之前，模型制作者可以依据材料的属性使用裁纸刀或锯对各层进行切割。

插入模型

一项任务或许包含若干个设计，而城市设计模型的制作通常是采用“插入”或者一组模型的方式以减少表现这一方案所需要的工作量。仅仅需要制作一个表现周边区域的模型，每一位参与设计的人员都会

图 7:
三维地貌模型

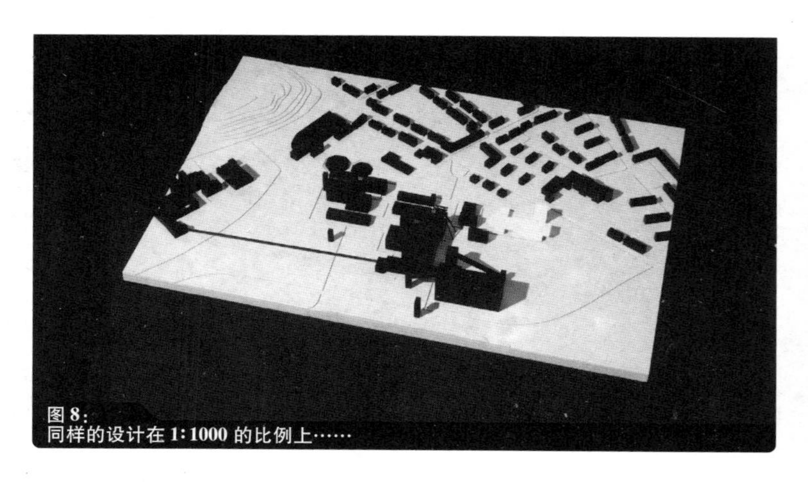

图 8：
同样的设计在 1:1000 的比例上……

图 9：
……这里的比例为 1:500

图 10：
……比例为 1:200

获得一块底板，在其上完成他或她自己的那一部分。这些特殊的部分会在城市设计模型中被省略掉，这样就使得插入的部分是可以替换的。

P16

建筑/房屋模型
（比例 1：200，1：100，1：50）

建筑模型是模拟建筑设计常用的说明性工具。大体量的建筑例如博物馆、学校与教堂在竞赛中通常采用 1∶200 或 1∶500 的比例来表现。除了三维的形式与体量，设计的多种多样的性格特征在这里与城市设计模型相比要扮演着更加重要的角色。

从图纸到模型

立面设计非常重要。在缩小的比例上，在建筑室外可以看到的建筑元素都吸引着人们的视线，其中包括：

— 立面；
— 表面，结构，特征，材料质感；
— 洞口与洞口的填充（窗户）；
— 斜面洞口，墙厚；
— 屋顶的形式与设计；
— 特殊细部例如屋顶的悬挑与女儿墙。

图 11：
在不同比例上的表现：1∶200 的建筑模型与 1∶50 的结构模型

建筑模型还可以表达室内空间与建筑结构的信息。例如，剖面模型（在不同的底板上将建筑分成两个部分）可以看到重要的室内空间。通过将可拆卸屋顶与其他的室内元素组合在一起，模型的制作者可以选择使用可拆卸的屋顶构件，这样就可以从模型上方看见内部效果。

根据设计的目的，模型也可以简化为只有结构或概念部分以实现更具说明性的效果。

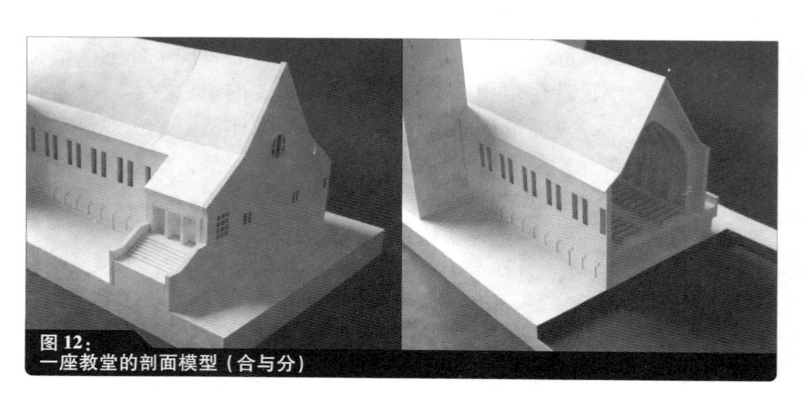

图 12：
一座教堂的剖面模型（合与分）

图 13：
一座教堂的剖面模型（室内空间的场景）

P19

室内模型（比例1∶20，1∶10，1∶5，1∶1）

当描述空间的时候，例如酒吧、礼拜堂或起居区域，通常宜于使用模型来对其真实的情况进行准确与详细的模拟。在这类模型中，抽象仅扮演着次要的角色。最重要的目的就是在小尺度上描绘真实的物体与材料。

怎样才能使各种表面、元素与物体在小尺度上再现，而仍然保留它们在整个结构中的效果呢？让我们用一座教堂的室内模型来解释这个过程。这个设计在将建成的建筑中仅使用了少许的几种材料。用简单的方式组合在一起，然而它们具有的特别品质对于空间气氛与主导理念来说都是至关重要的。真实建筑中的亚麻油毡地板在模型中也是用亚麻油毡复制的。复合木板漆成白色用以表现真实的白色石膏面层。玻璃部件则用玻璃制作。这样建造的结果则既是一个三维的模型，又是一种材料的集成。

图 14：
教堂室内模型——比例 1:10

图 15：
教堂室内模型——比例 1:10

模仿一种真实材料的最佳途径即是将这种材料用于模型之中。例如，模型的制作者可以在较大尺度的模型中用混凝土来表现精致的混凝土。他们仅仅需要建造一个缩小的框架结构，并将水泥与沙很好地混合，添充框架，将其压紧、干燥。结果将会非常富于感染力！在这个例子中，模型的制作过程是整个建造过程的缩影，不仅仅是对于即将建成建筑的模仿，而且也是对于建筑本身建造过程的模仿。

通常的原则是：那些对于真实建筑的效果有着最大影响的因素必须成功地反映在模型上。这样，模型将会非常有效地表现设计的空间。模型的制作者通常会用他们的室内模型获得令人惊讶的效果，欣赏者很难分辨出模型空间的照片与真实空间情景的照片之间的区别。

P21

细部模型 （**比例** 1：20，1：10，1：5，1：1）

细部模型不仅仅用于室内设计领域而且还应用于结构或技术模型（细部）。原则上，这些模型的比例可以做到 1:1，虽然这样做模型可能被称为“原型”会更为准确。模型的制作者首先要决定模型是否要

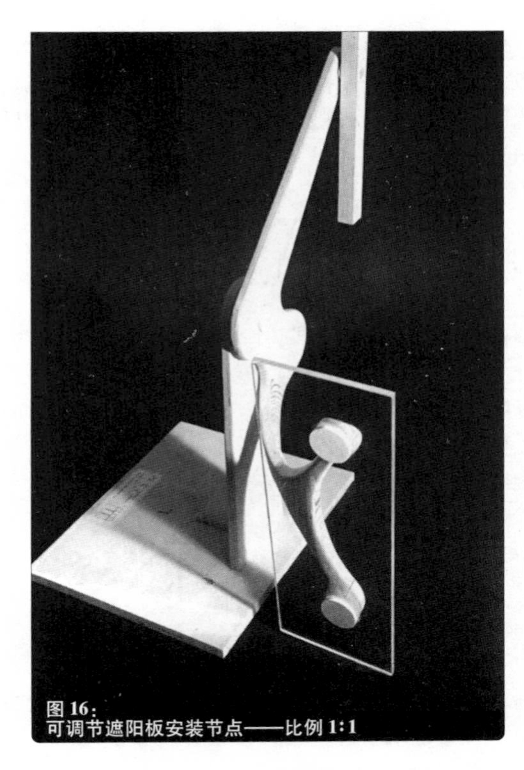

图 16：
可调节遮阳板安装节点——比例 1:1

图 17：
多层胶合板橱柜把手——比例 1:1

包括家具、灯具以及类似的细部，而且怎样做才能最有效果。当将旧的物体赋予新的功能的时候，创造是不受任何约束的。这些旧的物体可以作为比例为1:10的桌椅模型的原材料，或者也可以用于简单的建筑体块与抽象的建筑体量。

P23

设计与理念的发展

即便是模型也需要理念

模型制作是一个创作的过程，除了选择恰当的工具与材料外，它还需要设计的理念（见“设备、工具与技法”一章）。完成的建筑模型必须是一件美好的物品，向人们表达自己的内涵与设计。下面举的例子将会帮助学生们发展自己的设计方案：

— 色彩——单色的或多色的模型；

— 材料对比——区别材料的属性；

— 构成；

— 性质；

— 精细程度（“抽象”）。

P23

色彩与材料

单色模型

当决定了模型使用木材、卡纸板、金属或是塑料进行加工以后，模型的制作者必须确定他总共需要多少种不同的材料。一般来说，如果可以用不同的方法来修饰的话，一种材料就足够了。具有统一材质与色彩的表面的优点在于被表现的空间仍然是趣味焦点，不会被材料或者模型本身夺去注意力。制作一个单色模型是常用的手法。建筑竞赛中的多数模型被制作成“白色模型”，以石膏或聚苯乙烯塑料为材料：目的是将欣赏者的全部视线引向建筑或城市设计本身。木制模型通常要求只是用一种木材。不用说，如果其他元素没有削弱它的效果的话，所选择材料的美感将会以其最佳的效果呈现出来。然而如果区分不同的元素与组成部分很重要的话，模型的制作者可以使用涂料或清漆等等（见“材料”一章）。

区分不同的元素与表面是设计理念的一部分。在模型中准确地模拟各个构成要素的组合是一种经得起时间考验的模型制作方法。

一些建筑的实例：

— 光滑的石膏表面可以和粗糙质感的砖墙组合在一起；

— 致密的混凝土构件可以用来与轻质的、精致的木或钢结构进

图 18：
一个三色的建筑模型：基地是灰色的，建筑是煤黑色与自然的褐色（完整模型的比例是 1:200；细部是 1:50）

行对比；

— 透明的或半透明的元素（玻璃的建筑外表面）可以与不透明的部分组合在一起。

如果说对比是建筑理念的一部分的话，它通常会在模型中体现出来。在这种情况下，用简化的方法来表现材料以创造出更加抽象的感受是并不能够带来必要的清晰与精确的。然而如果将不同的材料组合在一起，模型的制作者则必须关注于什么是绝对必要的，以避免模型最终成为一个材料的大杂烩。

P24

构成与组成部分

模型的制作过程复制了在设计阶段已经发生了的那些步骤。无论是硬与软、暗与亮、重与轻、粗糙与精美，最重要的问题是材料应该怎样组合以充分表现它们的对比。大量的实验是必需的，这样才能作出最终的选择，并且确定材料的这种组合方式能够抓住模型的制作者所要传达的信息。只有在这些问题被解决以后，才可以开始真正的模型制作过程。

各组成部分之间的相互影响

我们怎样“构成”一个模型？目的是用模型制作的方法来强化设计的理念。

模型制作者必须首先关注以下问题：

— 怎样选择细节以使得建筑模型与整个模型表现的尺度成比例？

— 这个方案是要放置在整个表现模型的中央呢，还是有什么理由放弃这个原则？

✎

— 模型中各个单独的组成部分之间的关系是怎么样的（色彩、材料属性，比例）？

✎

小贴士：

底板的尺寸与形式对于模型的效果有着很大的影响。当你选择那些正方形、黄金分割、边长比为 1:2 的底板或将底板的形状与建筑平面做成一致的话，是几乎不可能犯错误的。另外一个选择就是基于建筑的形状选择底板的形状：如果选择一块长且狭窄的底板的话，将会强化狭长建筑的效果。

P25

抽象与精细程度

表现的模式与恰当的材料一道在最终结果的确定中扮演着关键性的角色。模型的所有组成部分都必须有着相同的精细程度。例如，如果新建建筑的模型仍旧是抽象的话，那么是不可能准确地表现基地与周边建筑的。

制作模型的比例将会决定抽象的程度。在模型制作的过程中，抽象意味着将对象简化到最基本的程度。非基本的元素可以被忽略掉或不予考虑。但是确切的来说什么是基本的要素呢？

抽象 = 诠释的自由

在这种情况下，对于所选择的抽象程度的思量是十分重要的：一个精准的、细致的模型必然是对合理的、经过精心设计的方案的描述。如果模型意味着提供大量的信息，欣赏者将会只有相对少的想像自由。真实结构的细致的缩微版本将会传达出设计意图中非常具体的理念，并且能够避免欣赏者对其自身的细节产生联想。有这样的经验，当与委托人或非专业人员打交道时，精细程度高是一种优势。模型越逼真，人们对于房屋或建筑的印象就会越清晰。

在模型制作当中可能达到怎样的精细程度呢？原则上说，惟一的限制是工艺上的可行性，或者在给定的条件下所可能够达到的程度。如果一扇窗子在选定的比例上太小，以至于不能用刀具切割出来，那么就省略它。

抽象程度较高的模型倾向于表达原则与理念，将会在以后的阶段充实细节。模型仍然是概念性的。建筑师通常喜欢各种极少主义的表达方式，这样就可以摆脱任何限制去想像并且可以对随后出现的建筑进行各种阐释。他们可以不必对自己负责。

配景

在模型制作的过程中使用树、人、车或其他的配景也必须与抽象的程度结合起来一并考虑。如果模型描述的不仅仅是建筑与基地，还包含人们在日常生活中所熟悉的元素，例如车、树、植物与人，那么通常会更加容易地向非专业观众传达具体的设计理念。“人体尺度”

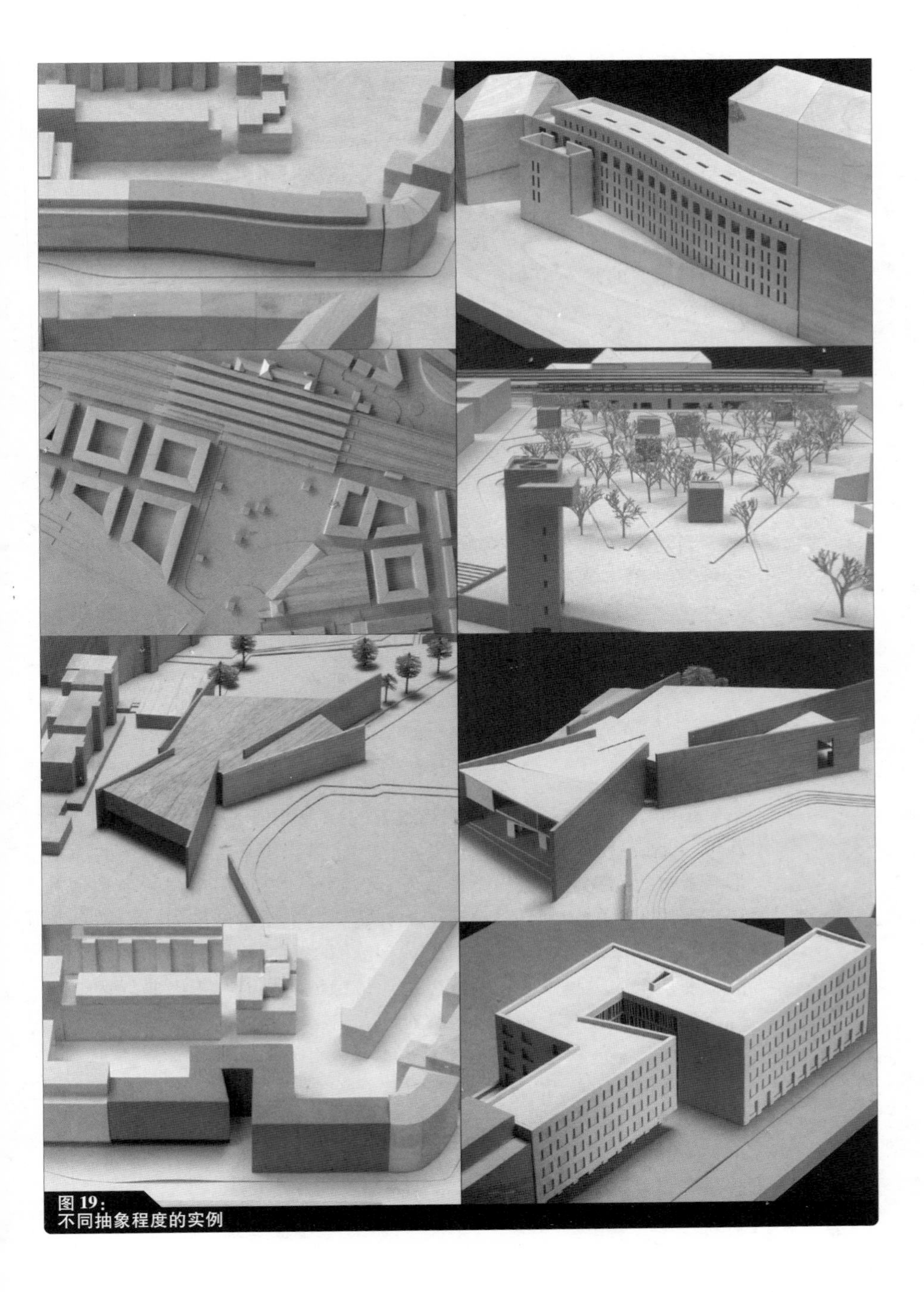

图 19：
不同抽象程度的实例

在现实生活中是最重要的准绳，但它并不总是在模型中体现出来（见“材料”一章中的“配景：树、人与车”）。

精准

模型制作通常与这样一些词联系在一起，如完美主义、精致的做工与准确的工艺。这些标准当制作表现模型时通常被证明是对的，其自身的精准确实能够保证模型的优良。但是对于模型来说这个标准甚至不是必备的条件。一个理念的清晰传达并不需要做很多的工作。比完美更重要的是模型制作者的“标志性”风格，这些我们在上文当中已经提及了，它给予模型以必需的表现力。板刻作为一种有创造力的方法，在缩微的尺度上，对于大的物体的呈现与精细的表现有着同样的效果。

P28

设备、工具与技法

当建造一个小的空间物体时，许多学生会感受到相同的热情，这种热情第一次将他们引领到作为一种职业的建筑学当中。另外一些学生会更喜欢铅笔与图纸，而不喜欢模型制作所需要的那些工具。无论他们的偏好是什么，所有的建筑学学生在他们的第一个学期都会要面临制作模型的要求。那么建造一个模型最简单的方法是什么？

P28

切割

最简单的方法是使用一把裁纸刀与一张纸板或卡纸板进行工作。裁纸刀是最基本的模型制作工具，因为它可以用来在许多材料的原始形状上进行切割。它很简单也很便宜，而且还有许多不同的种类可以用来完成不同的切割任务。我们推荐购买一把质量好的刀，而不是用那种在许多家庭装修店中所能够找到的简单的地毯刀。刀片一定要牢固，在切割的过程中一定不能晃动。最理想的是刀在手中能舒适地切合，这样就可以有力地控制与操纵切割刀了。刀具的选择就像选择一支好用的铅笔一样重要。

除了常用的可推拉的裁纸刀，在专业的商店里还可以找到其他的切割工具。手术刀，大家都知道它是外科手术的工具，可以用来在材料上作精细的切割，也可以在卡纸板上切割出非常小的窗子。

提示：

许多的模型制作者在使用裁纸刀时都有过痛苦的经历，尤其是在有时间限制的压力下工作的时候。为了将受伤的风险降至最小，你必须在使用工具的时候总是尽可能地小心，并且只在工作中，只在需要它们的时候使用它们。例如，你一定不能试图用一把裁纸刀去切割硬质木材，因为这样会有滑脱的危险。

小贴士：

当使用裁纸刀去切割卡纸板的时候，你握刀的位置应该与卡纸板的表面尽可能贴近，以获得精确、利索的切割效果。这个动作如果做得不恰当的话，材料会被撕破而且会增大受伤的危险，因为这样做锋利的刀刃会更容易滑脱。

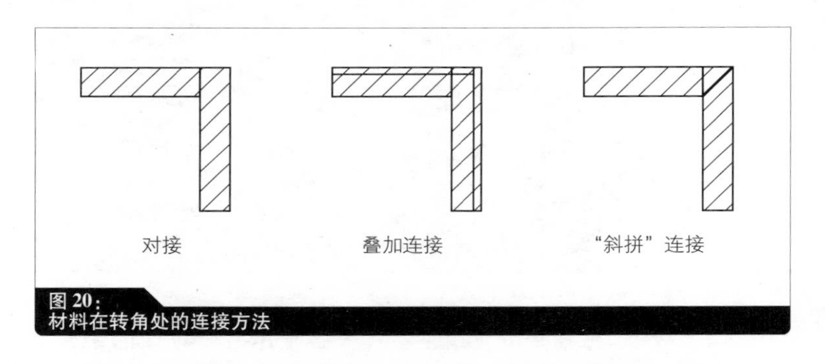

图 20：
材料在转角处的连接方法

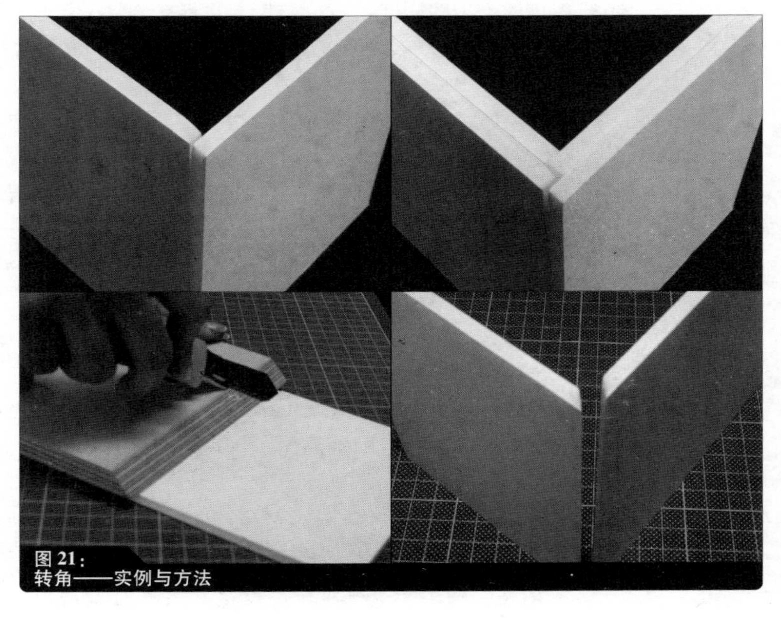

图 21：
转角——实例与方法

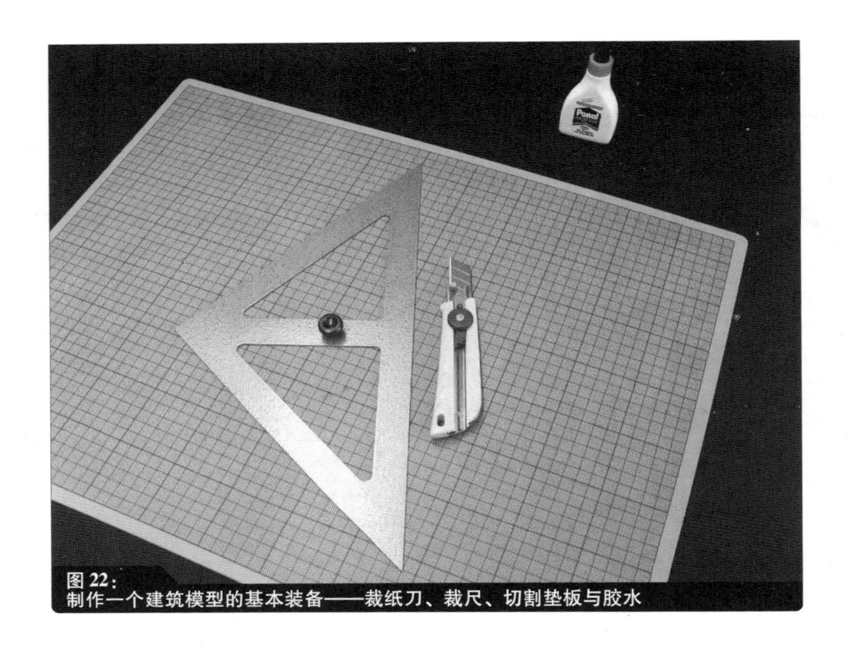

图 22：
制作一个建筑模型的基本装备——裁纸刀、裁尺、切割垫板与胶水

怎样来进行切割？首先必须用一支较淡的铅笔将需要的边界线标识出来；切割以后，这些淡淡的标识就看不见了。切割后的边缘要与材料的表面呈 90°角，这样各个部分可以对接或者粘结在一起。

在转角处，材料的边缘也可以切割成 45°角以形成斜拼连接。那种有着倾斜刀刃的专用裁纸刀可以用来完成这种切割。另一种方法是制作一块边缘是 45°角的木板，并将其用作模板，来引导标准的裁纸刀的刀刃走向。

另外还可以用一把金属边的尺子来界定刀的走向，以保证直线切割。还有一个重要的工具即是一块硬质的切割垫板，它不仅仅用来保护桌面，而且比起软质的基层来说使用它可以获得更好的切割质量。我们推荐使用那种塑胶的切割垫板。

这些仅仅是模型制作者用卡纸板或纸板做简单的工作模型时所需要的工具。但是当需要精细的加工材料的时候其他的一些工具就是必需的了，这些工具包括：

— 砂纸：用来打磨表面，完善切割边缘，将突出的小体块从切割掉的洞口中清除。如果将砂纸裹在一个坚固稳定的基座上的活，例如一个砂纸架，砂纸在需要打磨的地方前后移动将

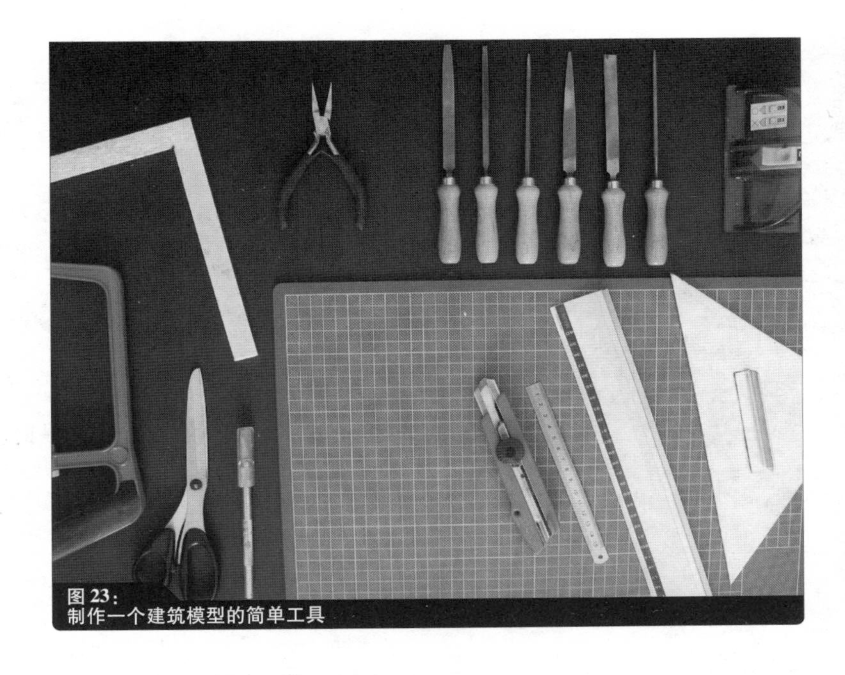
图 23：
制作一个建筑模型的简单工具

会变得非常轻松。

— 锉刀：非常适用于完善各种材料的转角与边缘。锉刀可用于加工木材与金属。通常来说加工塑胶与金属可以使用相同的工具——例如一把精良的金属锯在切割塑料时也可以做得很好。

— 钳子：可以提高人手的精度，帮助模型制作者在甚至最小的部分上进行操作。在小尺度的模型中会有许多部分只有几毫米的大小，用手指几乎不可能抓住它。

— 测量仪器（“滑动测径器”）：能够用来准确的测量直径与横断面。因为经过训练的眼睛可以在小比例的模型上察觉到甚至最微小尺寸的差异，因此在模型制作过程中精准的测量是一个非常重要的方面。将精度提高到 1mm 的 1/10，这个计量单位用普通的尺子是测量不出来的。

— 测量工具（尺子与量尺）：用于精确地测量长度与尺寸。

✎

小贴士：

在将模型粘结在一起之前，你必须首先找到一款最适合的胶水。大多数的胶水是在其包装中售卖的，没有为胶水准备恰当的工具。结果就是你或许会挤出过多的胶水，留下难看的污渍。

改善这种情况的一种方法是从药房买一只标准的带粗大导管的注射器，在其中注入选定的胶水。当粘贴透明的塑胶片时这种方法非常好用，因为在这类材料上的污渍是可见的、惹人厌烦的。

甚至还有一种更简单的办法，就是用牙刷或薄木片沾上双倍的胶水。

P32

粘结

基于所要粘结的材料，模型的制作者可选择的产品的范围很广。尽管万能胶普遍地适用于各种不同的材料，对于所有种类的木材、木质材料与卡纸板来说，白胶仍然是最理想的选择。

✎

表1：
胶粘剂总览

胶粘剂	属性	用途
万能胶（all-purpose glue）	通常是一种人工树脂的挥发性胶粘剂，它是透明的、黏稠的，涂在材料上时可拉成丝状，会与一些塑料发生反应（例如，聚苯乙烯泡沫），溶解它们的表面。通常在几分钟内就能干燥。有轻微的刺激性气味；不老化	可以适用于不同的材料（例如，卡纸板、木材、塑料、金属、玻璃、织物等等）。它可以将相同的或不同材料的组成部分粘结在一起。这种胶水不会使卡纸板发生变形，并且因为它的高度适应性，在模型制作的过程中有它就可以应付全部粘结工作了
白胶（木器胶）（white glue）	是一种白色、黏稠的胶粘剂，它通过吸收所要粘结材料中的水分来干燥。透明，干燥速度慢，这就意味着在涂完白胶后的一段时间内被粘结的表面是可以移动的。干燥更快的“快粘胶”（express glue）可以在3~5分钟内干燥	最理想的是用于木材、木质材料、卡纸板、纸板的粘结。粘连的部分必须压实在一起以保证有效的粘结。胶水当中的大量水分有可能改变材料的形状，或者导致卡纸板弯曲。不适用于塑料、金属与那些不吸收水分的材料
接触型胶粘剂（contact glue）	用于粘结材料的较大表面。这种胶要在需要粘结的两部分都进行涂抹，它是与自己粘结在一起的。涂完胶水以后，需要粘结的两部分在粘结在一起之前需要干燥几分钟，并且一定要用力压紧。接触型胶粘剂仅能在通风良好的空间中使用	在有高差地貌的模型中粘结材料的较大表面（例如卡纸板）的理想胶粘剂。因为胶水要在需要粘结的两部分都进行涂抹，而且空间必须通风良好，所以这是一种耗费时间的方法，但是它的优点是不会使材料变形。可以用于木材、卡纸板、大量的塑料、金属与陶瓷

续表

塑料胶粘剂(plastic-bonding adhesives)	一种胶水，通常是含溶剂成分的透明胶水。是专门为塑料设计的，单面涂抹的胶水。要趁胶水还未干的时候，尽快将需要粘结的部分粘在一起，材料的表面一定不能有灰尘或油脂	适用于多种热塑性塑料，包括聚苯乙烯、PVC与有机玻璃，但不可用于聚乙烯或聚丙烯塑料。也可用于粘结木材与卡纸板（见万能胶一项），而且对于塑料来说比万能胶的效果要好
强力胶水①	透明，能够非常快速干燥的胶水；黏度很大且不会滴落	是那些不能结合在一起以及需要瞬间粘结的材料之间的理想胶粘剂
喷胶（spray glue）	无色，防紫外线，不含氯氟烃，它装在喷雾瓶中，经涂抹后不改变颜色。因其较低的含水量，它不会渗入材料。喷胶仅可在室外环境或通风良好的室内空间中使用	是大面积粘结的理想胶粘剂，例如用于粘贴有高差地貌模型中的卡纸板。材料可能会有轻微的变形与弯曲，但通常不会褶皱。也可以用于将纸或卡纸板贴在不同的背景上，尤其是在需要大面积粘贴的时候
溶剂胶（solvent）	用于粘结塑料。这种溶剂胶溶解材料的表面，用力压紧以将两部分“焊接”在一起。例如：二氯甲烷(亚甲基的二氯化合物)，它与其他所有的有机溶剂胶一样对人体健康非常有害	用于将聚苯乙烯、有机玻璃或聚碳酸脂塑料黏合在一起；它能将热塑性塑料树脂溶解从而将两部分粘起来，并且不会有残留物产生。仅可在通风良好的空间中使用
橡胶胶水(rubber cement)	一种弹性胶粘剂，很容易除掉，而不留任何剩余物。如果在两部分都涂上这种胶水，将会产生永久性的粘连	可以用于多种材料，包括纸、卡纸板与塑料。是工作模型与组合构成的理想胶粘剂
双面胶带(double-sided tape)	是液态粘结媒介的基本替代品，因为其瞬间粘结的特性而用途广泛。相反，被粘结在一起的材料的属性却不会被湿气所影响	可以用于大面积粘贴，并且对于所有类型的材料都很适用，包括聚乙烯（PE）与聚丙烯（PP）等不能被其他胶粘剂粘在一起的材料。材料不会变形、弯曲或褶皱。不适于很小的点状粘结

P35

使用雕塑黏土

塑造、成型与浇注

制作工作模型的另外一种方法是用橡皮泥或其他的雕塑用黏土。

商店里通常卖的橡皮泥是一种灰绿色的模型制作材料，它的可塑性随着温度而变化。在室温下捏制橡皮泥是有困难的，但是一旦温度稍微提高，它就会变得更软更加容易制作（例如在温暖的散热器上）。还可以将橡皮泥放在罐子中进行加热，成为液态，这样就可以用油灰刀或刷子将其涂在某个表面上。

① 译者注：原著英文的用词是super glue，在中国对应的是强力胶或瞬间胶。

雕塑黏土可以很好地适应模型中的实验性工作。它非常适于用来研究城市空间，能帮助模型的制作者很快地建造出许多相同物体的不同版本。小型的结构可以非常快速与简便地用刀切割出来。

通过手或者工具，用黏土来制作，模型的制作者可以从手工工作及其结果中获得自由。结果或许并不是一件精美的模型，但是这种雕塑形式对于表达空间来说可以是一种非常富于表现力的方法。

石膏与可浇筑的混合物

石灰（石膏）是一种廉价的材料，非常易于加工。但是即便如此，制作石膏模型也是一个费时的过程，因为需要有两个步骤：

— 首先，要为随后产生的三维物体做一个精确的负模，并将液态的石膏注入其中。模具的精确程度决定了最终成果的品质。

— 当石膏干燥后，将浇筑物从模具中取出。

图 24：
橡皮泥模型

图 25：
城市发展模型，用白色石膏制作并喷涂了白色乳胶漆

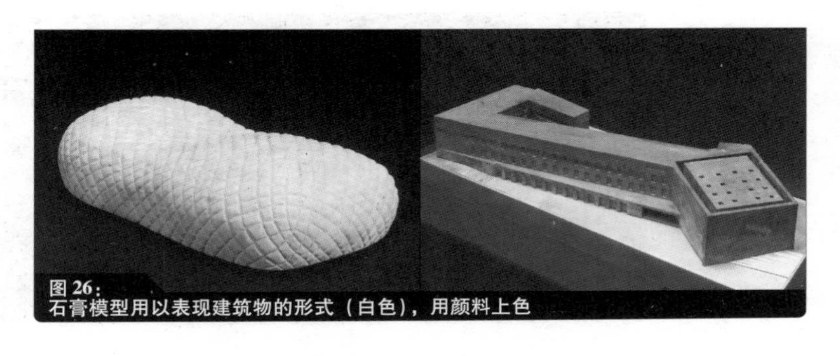

图 26：
石膏模型用以表现建筑物的形式（白色），用颜料上色

图 27：
混凝土模型用于表现实体的混凝土建筑

石膏模型经常用于建筑竞赛当中，尤其是城市设计的模型，因为在这种情况下，同样的基地模型必须作上几遍，而用同一个模具会很方便。可以在石膏中加入颜料或水彩，在其干燥后通过打磨或上色来表现色彩。这些模型不仅能创造出一种实体的、沉重的视觉印象，其本身也是实体的、沉重的。

不用说，除了石膏以外其他的可浇筑混合物也可以用来制作模型。如果需要表现美丽的混凝土外表面的话，模型的制作者也可以在模型中使用混凝土。

P37

模型工作室中的机器

使用机器

以上所讲述的简单的工具与方法对于易于加工的材料来说已经是足够了。然而，许多材料需要使用机器与专业的工具以获得理想的结果。

木工车间中的工具

许多职业的建筑模型制作者使用的工具与那些能够在木工车间中找到的工具相同：

图 28：
典型的木工车间中的手工工具

— 手锯；
— 锉刀、粗锉刀与砂磨块；
— 刨子；
— 凿子与锤子；
— 三角板。

锯切

当裁纸刀不能够满足需要的时候通常会使用电锯。虽然可以用刀来切割胶合板，但是木材这种材料通常还是需要用锯来加工。

除了那种常见的商用台式锯——有着圆形刀刃与长形工作台的那种锯子——较小尺寸的台式锯，叫做微型台式锯，也是（职业）模型制作者使用的工具。如果使用恰当的刀片，这种精良的电动工具可以用来割断或切削木材，同样也可以用来切割塑料。当制作木模型的时候手边正好有这类工具是非常有用的。使用锯架，可以横向切断木材的纤维。然而需要注意的是，学生应该有一定的训练与时间以熟悉这种锯子以及所要加工的材料的特殊性能。

小贴士：

将木材用于建筑模型的制作是非常吸引人的，这是因为这种材料的美感所致。即便如此，如果要使用机器的话，如台式锯或打磨机的使用，必须在职业工匠或模型制作者的监督下才可以进行，尤其是在最初的时候。

小贴士：

除了那些在五金商店以及 DIY 中心中所能找到的工具外，富于创造力的模型制作者通常会用非常简单的材料来制作自己的工具。很好的一个例证就是一把微型锉刀，它是由薄薄的一截方形钢管与细砂纸构成。也可以使用日常生活中的用品。例如，晒衣夹在粘结材料的时候能很好的起到夹紧的作用。

除了用来直线切割的台式锯以外，带锯也是用来切割曲线与自由形状的重要工具。这种机器在横向切割坚硬的木材时非常有用。曲线锯也用于切割自由形状与曲线。

刨平

刨子用于削减木材的厚度或者改变横截面的形式。通过刨削基地模型的表面，模型的制作者可以将标高的差别做得更明显。

钻孔

当进行连接的时候钻子十分有用，例如在模型的基地上使用榫钉以表现柱子或树木。在材料上钻孔可以将这些元素完美地连接在一起，并增加其稳定性。

图 29：
圆切割锯

图 30：
曲线锯

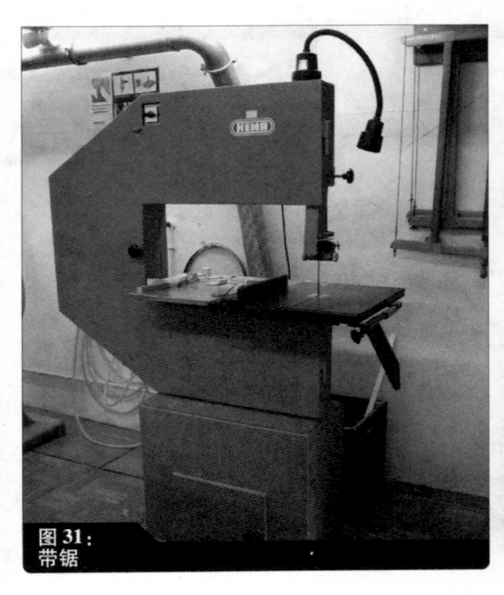

图 31：
带锯

图 32：
台式锯

图 33：
微型台式锯用于精细切割

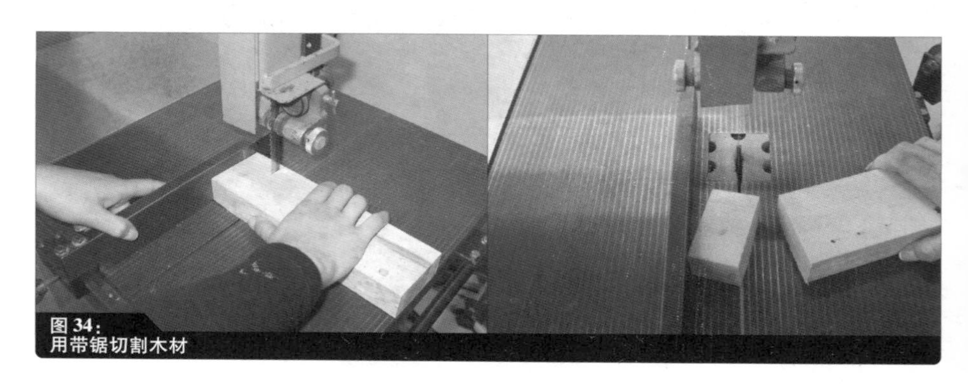
图 34：
用带锯切割木材

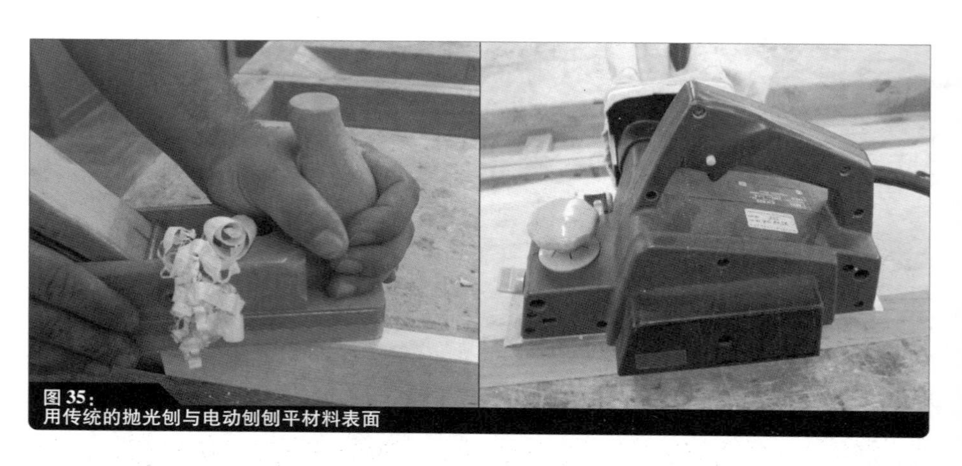
图 35：
用传统的抛光刨与电动刨刨平材料表面

图 36：
无导向系统的手持电钻

在使用钻子的时候，最重要的一件事就是正确地操控这件工具。在车间中，钻子通常被夹在钻架上，所以钻子进入材料的角度可以精确地调整，并且固定在恰当的位置上。用于模型制作的小型钻有着很小的卡盘，用这种钻子可以加工出直径小于 1mm 的孔洞。

铣削

模型的制作者可以使用铣床切开木材的表面。

模型的制作者可以在机器上接不同种类的铣钻头〉见图 37，以在浮雕模型的表面上切割出道路、河流、下沉区域及其他形态。除了上文所提及的用锯加工木材的各种可行的方法外，这种机器为木材加工提供了一种新的选择。

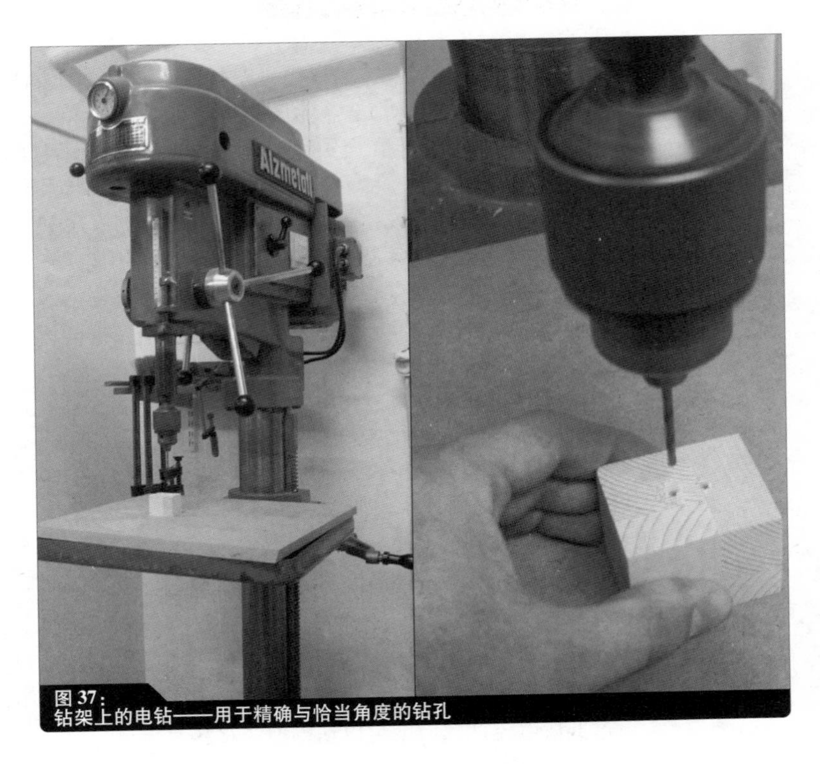

图 37：
钻架上的电钻——用于精确与恰当角度的钻孔

图 38：
铣床

打磨

打磨物体以获得最终的形式与表面有许多不同的方法。与用砂纸的手工打磨相比，电动工具会使得这项工作变得简单。在打磨的时候基于不同的材料选择恰当类型的砂纸——例如用于打磨金属或木材——是十分重要的。粗糙的打磨可以选择使用粗糙颗粒的砂纸，而细小颗粒的砂纸则应该在后期使用。电动的圆盘磨光机是一种非常有用的工具。旋转的打磨盘是由平展的工作表面构成，这样大面积的材料表面也可以被精细打磨。就像台式锯的使用一样，也会使用小型的打磨机以完成精细与准确的制作工艺，这对于精准的模型制作来说是十分必须的。

除了平板的圆盘打磨机外，还有一些其他的工具有着圆形的打磨表面，用于磨去突出的形体。它们包括砂带磨光机与往复式砂带磨床，它们是用手来牵引在需要加工的表面上进行打磨的。

P43

电热丝切割机

电热丝切割机（聚苯乙烯切割机）是建筑学的学生们经常使用的工具。它能够对很厚的聚苯乙烯泡沫进行简单、快速并且准确的切割，而且也可以用来制作城市设计模型中的单体以及规划过程中的体块模型。电热丝切割机的组成部分包括切割平台与细电阻丝，在其中通过低伏电流加热后就可以切割泡沫材料。

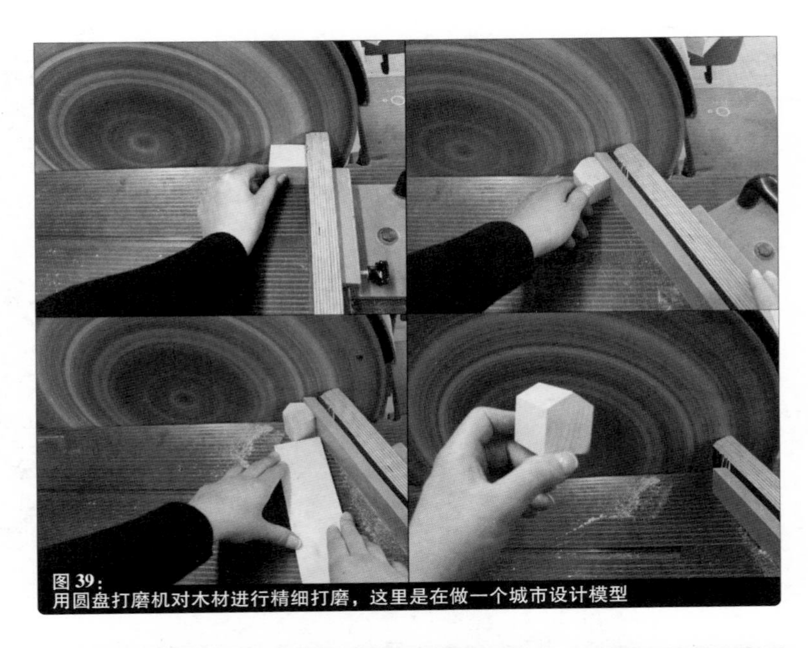
图 39：
用圆盘打磨机对木材进行精细打磨，这里是在做一个城市设计模型

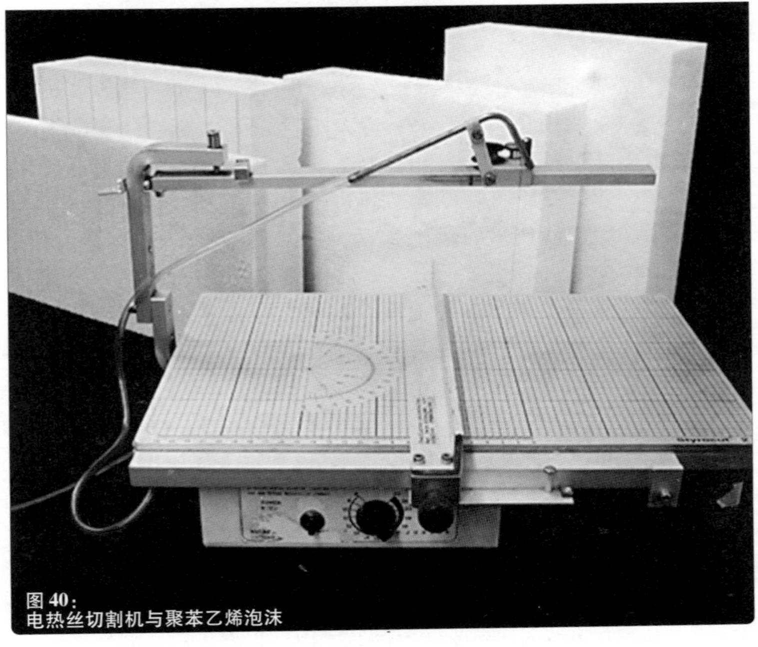
图 40：
电热丝切割机与聚苯乙烯泡沫

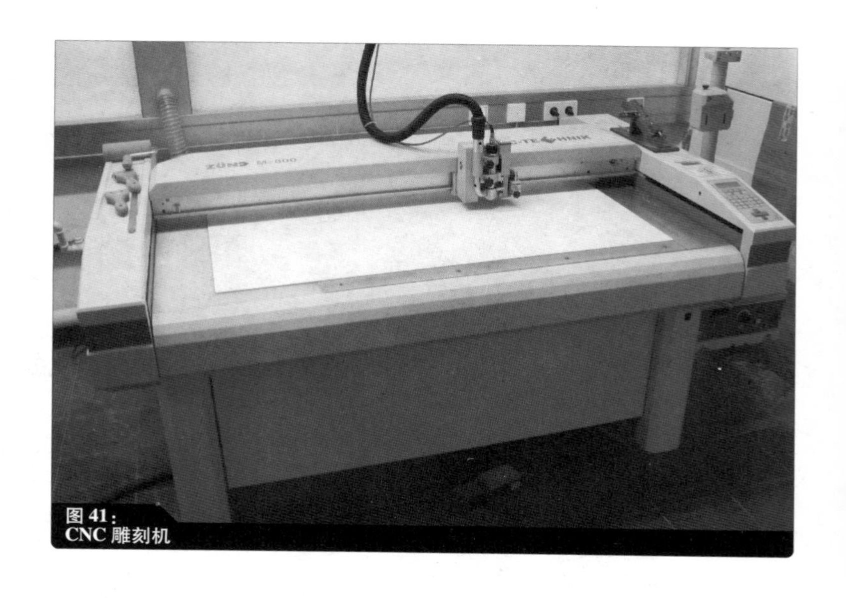

图 41：
CNC 雕刻机

P43

电脑雕刻机

计算机的使用已经改变了建筑学领域，尤其是在表现复杂的有机形体与结构方面。CAD 软件与 CNC（计算机数字控制）雕刻机——一种用于建筑模型制作的现代的数字工具——在这种变化中扮演着最重要的角色。严格来说，使用 CNC 雕刻机的三维表现所涉及的也只不过是用 CAD 软件进行二维的制图。就像草图绘制者的工具被数字化了，使用 CNC 雕刻机进行建筑模型制作也是如此。

从技术的角度来说，CNC 雕刻机，通常是由计算机对其在三个空间轴（X、Y 与 Z 轴）方向上进行操控的，它能够制作出高度精确与完美的模型。就像数字绘图仪与使用手工方法相比能够绘制出更加出色与更加准确的线条图一样，切割与雕刻机可以从多种适合的材料中切割出在图形上或空间上都很复杂的形式。另外，还可以随心所欲地经常制作模型，而保证个别的部分永远都相同。因此这种机器开启了一个非常广阔的领域，为以往所没有用到过的图形表现提供了可能性。然而一些充满激情的模型制作者却感觉那些用 CNC 雕刻机制作出来的作品中没有了他们自己的性格。

除了传统的模型制作工艺外，现在的许多大学也将这些新的数字方法纳入建筑学学习的范围中。根据机器类型的不同，全部必需的模型制作数据（取决于以矢量形式表达的地形或建筑有关信息，还包括立面的结构）都必须在所有的雕刻工作开始之前准备好。这些数据能够使机器从材料中切割出正确的形状。如果雕刻机的软件不能够恰当地翻译 CAD 文件（例如，因为过时的文件版本）的话，问题就会发生了。这就意味着所有的东西都必须重新画一遍。

提示：

基于设计，CNC 雕刻机可以很大程度上推进模型制作的进程，尤其是在再现有许多洞口的立面的时候。大学、职业的建筑模型制作者与公司在有限的时间内制作模具或原型，通常倾向于使用这种机器。

小贴士：

在雕刻的过程中，包含着二维图纸到三维空间形体的数字转换过程，在图纸当中的所有错误将会变得显而易见。因此注意以下几点是非常重要的：

— 线必须总是整洁的，而且在其端点要精确连接；
— 不能使用双线；
— 线被赋予色彩的方式必须满足雕刻机可以定义线的右侧与左侧的要求。

为了使用 CNC 雕刻机获得最佳的结果，模型制作者必须花费足够的时间用于检验这种现代工艺的运转情况。去注意切割头在预定的线上是向左或向右移动，还是径直地沿着线本身移动是非常重要的。

并不是所有的材料都适合使用 CNC 雕刻机进行加工。铝、黄铜、铁与不锈钢在薄片状态下是可以使用的（通常是 3 ~ 5mm 的厚度）。质量好（模制）的有机玻璃的最大加工厚度是 10mm，塑料，如聚苯乙烯，也是如此。木材如胶合板（举例来说，桦木与多层胶合板以及 MDF 即中密度纤维板）也都可以使用这种机器来加工。

P46

材料

何种模型用何种材料?

严格地说，模型只不过是已经存在的环境或一座未来建筑的小尺度版本。在模型制作的实践中，这就意味着真实的表面经过抽象的过程被转化为模型。然而必须尽量保持材料的特有属性与其可能的效果，因为最终一座建筑的外观主要来自其建成材料的综合效果。材料是暗哑的或是发光的、是粗糙的或是光滑的、是沉重实体的或是轻巧易损的。这样就提出了一个问题，即为了最佳的模拟真实，模型中应该使用什么材料。在这点上，模型制作与真实的建造是一样的。各种各样的材料都可以使用。其中一些材料用于建筑模型制作已经很久了，被认为是“经典”的材料；一些是新材料；还有一些最初用于其他的用途。创造特别的效果必须具有特殊的性能。玻璃可以用透明材料来模拟；水用反射材料模拟；而墙可以用层状材料来模拟，这种材料由小的模数组成，有着规则或不规则纹理的表面。

以下，分类描述了材料的使用与准备方法。可以依据材料的特定来源将其分类：

— 纸、纸板与卡纸板；
— 木材与木质产品；
— 金属；
— 塑料。

其他的重要材料如下：

— 颜料与清漆；
— 石膏与黏土；
— 橡皮泥与模型黏土；
— 模型制作中的预制部分，如人、车辆与其他的配景。

P47

纸、纸板与卡纸板

在市场中有各式各样的这类材料，而且用它进行建筑模型的制作，对于创造力的发挥不会有任何限制。许多产品，包括灰色纸板，实际上是为了包装工业开发出来的，然而在模型制作者当中却非常受欢迎。

表2（第115页）对于最常见类型的纸板进行了总结。纸，包括彩色纸或描图纸，也是模型制作的有用材料。卡纸板，在售卖当中使用的名字有双面白色纸板（mill blank）、色卡、蜡光板或克洛麦鲁斯板（Chromalux board），可以像纸与纸板一样用于表现结构表面的质感。

提示：

多数的方案都可以用这些材料轻松地制作出来。除了在加工方面很方便外，它们还有如下的优点，既便宜而且在许多地方都可以找得到。

图 42：
模型制作纸板——不同色彩与质感的选择

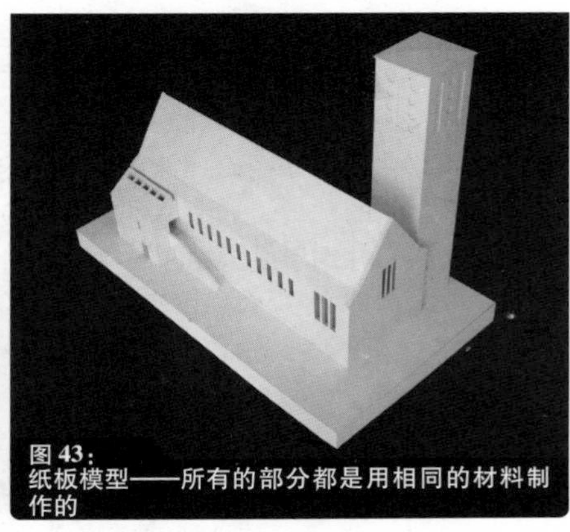
图 43：
纸板模型——所有的部分都是用相同的材料制作的

表 2：
纸板——使用与特征

材料	属性	用途	加工方法
芬兰木材纸浆板（也被称为芬兰板）	用木材纤维制作；米色、木材本色；当暴露于阳光下的时候颜色变深（黄色）；有着光滑或者粗糙的质感；可以找到的材料的厚度大约是 1.0～3.0mm	通常用于地形的阶层与建筑，以及模拟白天效果的室内模型（因为它有着光亮的表面）	容易切割；用白胶或万能胶就可以轻易地将其粘结在一起
灰色纸板	100%可循环利用；温暖的灰色调；有着光滑或者粗糙的质感；可以找到的材料的厚度大约是 0.5～4.0mm，随着制造商的不同而不同	通常用于地形的阶层与建筑	容易切割；用白胶或万能胶就可以轻易地将其粘结在一起；表面可以涂清漆或颜料
丝网板	木质纤维板；拼贴的纹理；可以找到的材料的厚度大约是 1.0～3.0mm	是空间模型的理想材料（比例 1:50）；因为它是白色的表面，因此可以用来模仿日光的效果	容易切割；用白胶或万能胶就可以轻易地将其粘合在一起。表面可以涂丙烯酸颜料或刷上丝网印刷漆

相片卡纸

相片卡纸，与淡彩纸一样，非常适于表现不同色彩的表面。在许多制造商那里都可以获得这种材料，它具有多种多样的色彩，而且因厚度的原因它对于模型制作来说是一种稳定的材料。

蜡光板

蜡光板的名字与材料的特性之间有一些歧义。因其有光泽的外观，这类卡纸板总是适用于表现反射表面与镜面。除了具有光泽之外，众多不同的色彩也是它的优点之一。

皱纹卡纸板/波形板

那些便宜且容易获得的材料对于制作工作模型来说总是非常实用的。皱纹卡纸主要是为了包装而生产与使用的，而且它可以被“回收”并用于模型制作。这种材料在包装方面的优点也表现在模型制作上：

— 它可以很容易地用裁纸刀进行切割；
— 6mm 的厚度使其非常易于加工；
— 这种材料的结构（皱纹卡纸板的核心，两侧覆盖着光滑的卡纸板）意味着虽然有轻微的荷载，它仍然能够保持其原有形状，并仍旧坚挺；
— 它也适用于有高差的模型。

图 44：
灰色纸板制作的模型

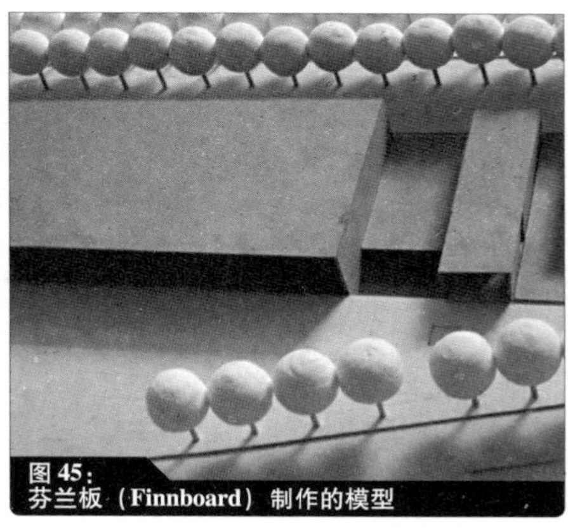

图 45：
芬兰板（Finnboard）制作的模型

图 46：
白色表面的纸板模型

图 47：
染成白色的纸板（克洛麦鲁斯板）模型

图 48：
用皱纹卡纸制作的模型，表现在基地表面（地平面）以下的建筑扩建

图 49：
由灰色的卡纸板（基地）与桦木胶合板构成

图 50：
用某种木材制作的城市设计模型（均质的）

图 51：
用不同种类的木材制作的城市设计模型（对比）

图 52：
木材：胶合板与条板

P51

木材与木质材料

木材用于制作建筑模型要比任何其他的“建筑材料”都要久远。甚至米开朗琪罗（Michelangelo）都曾使用菩提树的木材制作圣彼得大教堂的模型。木材与纸板相比是更能够进行深度加工的材料；而且效果也更加富于表现力。木材首先用于表现模型。作为一种天然的原材料，木材有着其自身的美感，这种美感可以独立于建筑模型的形式、设计与结果来表达。细微的色彩差别与纹理的结构使木材的外观成为模型中“有生命的”组成部分。

有两种基本的类型：自然生长与干燥的木材，与工厂加工出来的木质产品。两者都可用于模型制作。

小贴士：

在贮木场以及木工车间与木材加工厂中可以获得各种各样品质优良的木材。要概括来自全世界的各种类型的木材几乎是不可能的，但是专业人员通常会非常高兴地为建筑师与学生选择恰当的模型制作木材提出建议。而且还可以观察木匠与家具师的工作，向他们学习怎样加工木材。

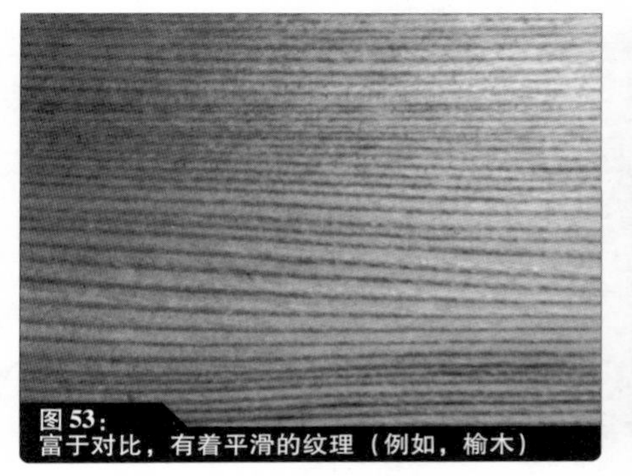

图 53：
富于对比，有着平滑的纹理（例如，榆木）

图 54：
深色与细微的纹理（例如，鸡翅木）

图 55：
大木纹，黄褐色木材（例如，橡木）

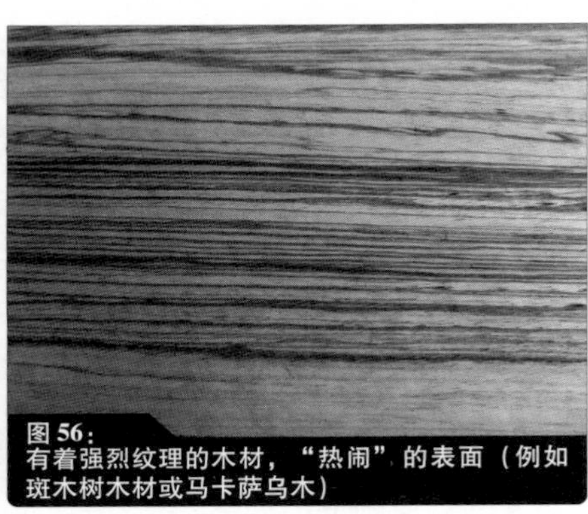

图 56：
有着强烈纹理的木材，“热闹”的表面（例如斑木树木材或马卡萨乌木）

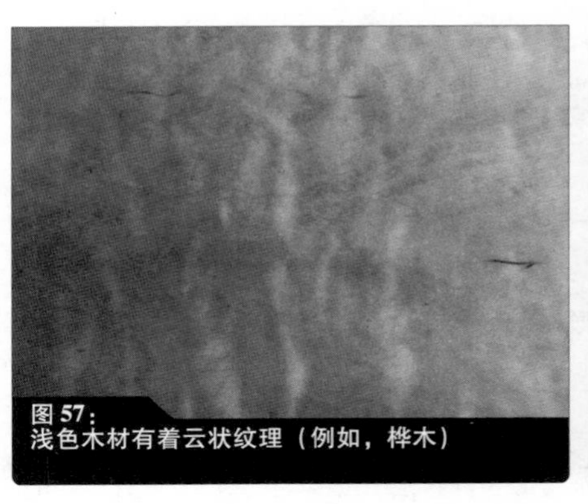

图 57：
浅色木材有着云状纹理（例如，桦木）

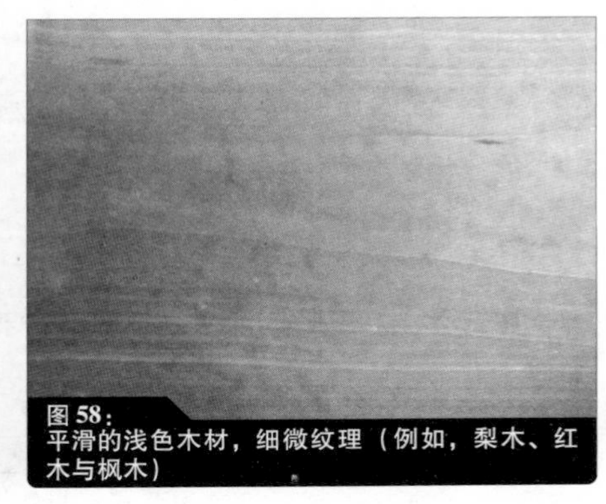

图 58：
平滑的浅色木材，细微纹理（例如，梨木、红木与枫木）

表 3：
木材——使用与特征

树种	属性	用途	加工方法
北非橄榄木	柔软，质轻的阔叶木，低强度，浅色，淡黄色，表面纹理较弱	用于制作船模的甲板。薄板条用以表现木材表面（凸凹的纹理）	薄板条可以使用裁纸刀（沿纹理方向）轻易地切割。使用白胶，或者大面积地使用喷胶可以将它们很好地粘结在一起
西印度轻木	所有木材商品中最轻质的。有微弱光泽，发白的表面，柔软均质的结构	尤其适于制作飞机模型。薄板条用以表现木材表面；用其外形轮廓表现建筑	可以使用裁纸刀或锯轻易地切割。这种木材很容易沿纹理裂开；用白胶可以非常容易地将其粘结在一起
山毛榉木	结实的硬木，细微的、均匀的纹理结构，淡棕红色	适用于所有类型的模型：木块用于表现城市发展；薄板用于建筑的表面与模型	使用标准的木材加工工具。可以锯与打磨。用白胶可以非常容易地将其粘结在一起
菩提树木材（酸橙树）	松脆的纤维，软质阔叶木，清晰的表面，淡黄颜色	在模型制作中最常用的木材之一。它适用于多种用途	依据尺度，菩提树木材可以非常容易地用裁纸刀或者锯子对其进行加工
桃花心木木材	非常坚硬的热带木材，有些许光泽的表面，暗红褐色	用来与所有的浅色木材相对比，是模型制作材料	因为它很坚硬，所以只能够用锯或打磨工具进行加工
枫木	柔软的阔叶木，天然的黄白色；细小、明显的纹理图案	可以用于所有种类的模型，木块可以用于城市规划模型，薄板用于表面与建筑模型	使用标准的木材加工工具，锯或打磨。用白胶可以非常容易地将其粘结在一起
梨木	平滑纹理的硬木，浅红棕色，精细的表面	适用于所有种类的模型：木块可以用于表现城市发展设计方案，薄板用于表面与建筑模型	使用标准的木材加工工具，锯或打磨。用白胶可以非常容易地将其粘结在一起
松木	长纤维的软木，典型的、显著的表面纹理，黄色	它在空间上的稳定性，使得它非常适于制作框架结构的木制的凸凹断面与构造模型	依据尺度，松木可以非常容易地用裁纸刀或者锯子对其进行加工
胡桃木	依据不同的产地：有着精细或粗糙的纹理，深棕色	用于与所有的浅色木材形成对比，是模型制作材料。有着非常精美的外观	使用标准的木材加工工具，锯或打磨。用白胶可以非常容易地将其粘结在一起

木材与树种

当模型的制作者使用天然木材的时候，他必须使用一种也适用于小尺度模型的木材。粗糙纹理的木材，有着显著的条纹，例如马卡萨乌木或斑木树木材是不适用的（见图56）。木材的表面看起来应该光滑平整。表3中所提及的木材是强力推荐的。

木质材料

木质材料，常用的是镶板，它是由木材加工厂利用废弃木材生产出来的。在建筑工业中，它们被应用于家具与室内的装修，也用于建筑模型的制作。这种板材也可用于制作底盘，例如模型的负重底板，并且可以用来制作大尺度的室内模型。

表4：
木质材料——使用与特征

木质材料	属性	用途	加工方法
刨花板	价钱合理的木质板材，由木材的刨花加工而成，粗糙表面，材料厚度：6.0～22.0mm	用于制作底板，通过表面处理来表现表面的凸凹与结构	可以像木材一样加工（锯或打磨），用白胶可以非常容易地将其粘结在一起。如果暴露在湿度很大的环境下会弯曲变形
中密度纤维板（MDF）	坚硬的、高密度的木质纤维板，均匀、光滑的板材纹理，很好的空间稳定性，平整表面，“自然的棕色”也可以染成各种不同的颜色，包括黑色	用于制作底板、浮雕模型与完整的、大比例（1:50）建筑模型	可以像木材一样加工，非常容易粘结，表面可以通过油漆或着色进行处理
胶合板（桦木、山毛榉木、白杨木）	胶压木质板材，有很多层，使用者集中木材之一制成，可见纹理，与所使用的木材色彩相同	用于制作浮雕模型与建筑模型。是完全由木材制作的模型的另外一种选择，因为有着较小横断面的胶合板也是非常稳定的	用裁纸刀或锯可以非常容易地对其进行加工，在某种情况下也会用到打磨机（桦木）。可以用白胶非常容易地将其粘结在一起
芯板材	用胶粘结木材制成的胶合板；可见纹理，与所使用的木材色彩相同	适用于制作底板与基地	可以像木材一样加工。因为它的构成（棒材胶合）通常可以仅在一个方向上承担荷载

图 59：
木质板材：桦木、胶合板、刨花板、中密度纤维板

图 60：
1:50 的模型，用中密度纤维板制作

图 61：
用中密度纤维板制作（漆成黑色）的基地，用菩提树的木材制作的设计方案建筑模型（在材料与色彩上进行对比）

图 62：
用木材制作的模型轮廓以表现结构框架，而中密度纤维板用于制作底板

P56

金属

当表现金属的时候必须特别注意，因为它们特殊的性质不能够用其他的材料忠实地模拟出来。如果想要使用材料来传达理念背后的美感，它将也会影响到模型所创造出来的效果。拉丝钢的柱子与拉接杆件非常适于表现纤细的金属轮廓，在这种情况下用其他的材料来表现这种特殊的性质将会导致失败。

金属板与金属型材都可以用于建筑模型。光滑的金属板，厚度大约是 0.2 ~4mm 就可以用来加工。还有一些带纹理的金属板，如带沟槽的金属板、波状金属板与网纹钢板；以及带圆洞、方洞或拉伸穿孔的金属板，还有金属网。

为了真实地表现钢铁型材，也可以买一些微型的实心型材，圆管，T 形、L 形、I 形型材，方形以及矩形型材。

表 5：
木质材料——使用与特征

金属	属性	用途	加工方法
铝	银白色，遇空气与水不锈蚀，因为它有一层厚的、不透明的氧化层，这是它在被使用的过程中形成的，不为磁石吸引	铝板与型材可以用于表现金属的建筑构件，例如，将波形金属板用于屋顶	不可以焊接。用胶粘结（万能胶），容易抛光，可以锻造成型
黄铜	铜与锌的合金。红色至浅红色，取决于所用铜的属性。通过使用较高比例的锌，可以生产出金色的黄铜	黄铜板可以用于模拟发光的金色表面；型材可以用于模型的承重部分（桁架与柱子）	可以焊接。结合牢固。依据材料的强度不同，可以使用金属工具对其进行加工。抛光良好
铜	惟一的红色金属，遇空气氧化，首先变成红色然后变成绿色	铜板与型材在模型中用于表现铜的质感	可以焊接。胶粘效果良好。依据材料的强度不同，可以使用金属工具对其进行加工。抛光良好
铁与钢	暗银色，与潮湿及氧气接触会锈蚀，产生红棕色铁锈；为磁石吸引	钢板可以用于模仿金属表面；型材可以用于模型的承重部分（桁架与柱子）	可以焊接与熔接，也可以用万能胶将其粘合在一起。材料需要防腐保护（电镀）或随后刷漆。可以使用剪刀或金属切割锯进行加工
镍银	铜、锌与镍组成的合金，银色表面，有着良好的抗空气腐蚀能力	镍银板适于表现金属质感的与发光的建筑构件	可以焊接；胶粘效果良好。是无屑加工（锻造成型）的理想材料
不锈钢（V2A 型）	银灰色，光滑，表面精美，不为磁石吸引。这种材料经过加工，不会生锈	适于暴露于潮湿的环境中使用（例如，室外）	用胶粘合（万能胶）

图 63：
建筑模型，立面是非常精美的穿孔金属板

P59

塑料

我们有许多不同种类的塑料产品，用通常的语汇很难描述它们。大多数的塑料都是具有延展性的人工合成材料，由大分子化合物组成。炭（一种有机材料）是其主要的组成成分之一。所有的塑料都容易加工，并且有着高度的精准性。它们轻质的特征、稳定的组成使得它们在那些其他材料所不可能的领域中非常适用。

聚苯乙烯模型

在模型制作中最常用的塑料是聚苯乙烯（PS）。大规模生产的聚苯乙烯——与聚丙烯（PP）和聚氯乙烯（PVC）一样——价钱不高，而且用途广泛。许多建筑师与模型的制作者只使用聚苯乙烯进行模型制作。结果就是，在建筑模型制作领域中发展出一类特殊的设计与表现方法。聚苯乙烯是白色的，光滑的。它是精准加工与制作精美形式的理想材料。完全用聚苯乙烯制作的模型可以表现所需要的各种抽象形式，也可以用于制作纯净、简单的三维建筑实例。

使用塑料可以获得几乎完美的效果，这种特征将它与其他的材料区分开来。可以在 1mm 以内的精度对塑料进行精确的加工，举例来说，这在制作城市规划模型时这就是一个非常大的优势。另外，它还可以用来模仿透明构件，例如用薄的透明 PVC 板可以模拟玻璃。塑

图 64：
塑料——为模型制作提供了多种选择

料是便宜的模型制作材料。聚苯乙烯是建筑师在竞赛项目与一般的模型制作中最常使用的材料。

模型制作中的亚克力

除了聚苯乙烯外，亚克力玻璃（化学名称是：聚甲基丙烯酸甲酯，PMMA；也被称为有机玻璃与树脂玻璃）则是最广泛用于建筑模型制作的塑料制品。作为一种热塑性塑料，它表现出良好的热学展延性。另外，可以通过许多不同的加工方法非常理想地表现玻璃窗格、玻璃以及透明的建筑构件（例如，城市规划模型）。它的表面可以被修改。通过磨碎亚克力玻璃（所谓的“湿磨法”，用细小颗粒的砂纸，例如颗粒 600），可以制作粗糙的纤维纹理。用锋利的裁纸刀刃在亚克力玻璃上铣削、开槽、刻划，可以做出栅格的纹理与图案。

提示：

在模型制作的过程中，PS（聚苯乙烯）硬质泡沫最终的成品经常会变成不同的颜色。不幸的是，一些涂料的溶剂——尤其是如果喷涂的话——会与塑料化合物产生反应，而且甚至能够溶解它。因此可取的方法是首先进行实验。

小贴士：

亚克力玻璃通常会很贵，而且很难在专业经销商那里获得这种材料。因此值得与加工工厂或制造商直接联系，索取废弃材料或样品。

我们有多种多样的塑料与塑料产品。在模型制作中所使用的塑料的种类也要远远多于这里所描述的。选择使用塑料的惟一准则就是形式、想要的色彩以及材料的表面属性。在可以用于模型制作的，以板、箔以及型材的形式出现的塑料产品当中，需要特别提及的是聚酯与聚乙烯（PE 塑料）。

除了以上所提及的用于制作建筑模型的“原材料”，许多其他的产品——如涂料与清漆——也是不可或缺的模型制作材料。

表 6：
塑料——使用与特征

塑料	属性	用途	加工方法
聚苯乙烯（PS）：作为硬质塑料，“硬”塑料	抗振，硬质塑料。不光滑的、白色、不透明板材用于模型制作。聚苯乙烯不能防紫外线。材料厚度 0.3~5.0mm	通用：用于模型制作的所有领域	很容易用裁纸刀切割，其表面易于抛光，并且非常容易削磨。它们可以用溶剂胶、特殊的聚苯乙烯胶水或接触型胶粘剂轻易地粘合在一起。易于刷涂涂料或清漆
聚苯乙烯硬质泡沫（例如，聚苯乙烯泡沫塑料）	一种多孔材料，制成板状或块状。不抗振。其表面受挤压很容易下陷。我们可以得到不同颜色的聚苯乙烯泡沫塑料，这都取决于制造商	用于建筑与城市规划模型	非常容易用电热丝切割机切割，用裁纸刀切割或雕刻。也可以被抛光与上色
聚丙烯（PP）	抗热、硬质、抗拉扯塑料。在模型制作中，这种材料通常是以薄的、透明或不透明的箔片形式被使用的。它有着不能被划坏的表面，具有紫外线稳定性，而且可以获得多种不同厚度的聚丙烯	作为一种半透明箔片，它非常适于模仿毛玻璃表面与灯光设计	很容易用裁纸刀切割。它可以被弯曲、折叠、挖槽、贴边、铸造以及打孔。在粘合之前必须对其进行预先处理，例如，使用聚合物的底漆
聚氯乙烯（PVC）	有透明的或不透明的，也有不同的材料厚度，这都取决于制造商的加工	透明的箔片使用在模型中可以很好地模拟玻璃，也可以在许多不同的领域中作为箔片使用	很容易用裁纸刀切割。它可以被钻、铣削、扭转，PVC 塑料的表面可以用标准的塑料胶或接触型胶粘剂粘合在一起

续表

聚碳酸酯（PC）	非常坚韧，耐冲击塑料，耐候，表面精致，透明或乳白色半透明塑料箔片	透明的箔片使用在模型中可以很好的模拟玻璃，也可以在许多不同的情况下作为箔片使用。它的表面与PVC塑料相比，通常要更加的精致与光滑	很容易用裁纸刀切割。厚板可以被刻划或折断。PC塑料的表面可以用溶剂胶或接触型胶粘剂粘合在一起，并且没有残留物
亚克力玻璃（PMMA）	非常透明、明亮，有很好的光学性能，与玻璃相似。耐候塑料，可以获得透明的、半透明的或不透明形式的亚克力玻璃	作为一种透明材料可以用于表现玻璃或水体	箔片可以轻易地用裁纸刀切割，较厚的材料必须折断或锯断；在打磨与抛光之前需要打湿。可以用溶剂胶、接触型胶粘剂或亚克力玻璃专用胶轻易地粘合

图65：
单色聚苯乙烯模型，用乳胶漆喷涂成白色

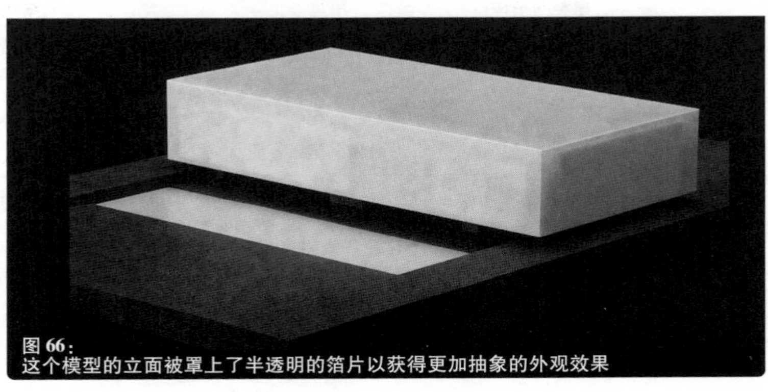

图66：
这个模型的立面被罩上了半透明的箔片以获得更加抽象的外观效果

图 67：
城市规划模型，其建筑是用硬质的拉伸聚苯乙烯制成

P64

涂料与清漆

材料的表面并不总是在完成的模型中所能够看到的那个表面。涂层、薄膜与面漆可以用于多种不同的基层以修改原始的设计。这些产品也提高了材料表面的抗性（绝缘、抗紫外线、防水表面、防止变黄）。

小贴士：

当你使用涂料、清漆或其他的表面涂层的时候，应该首先做实验来检验一下它们的效果。使用样本与测试材料，你可以获得产品属性以及使用它们的不同方法的直观感受。实验也能够非常有趣。在自然光线下能够对色彩与对比作出最好的判断。

表 7：
涂料与清漆——使用领域

涂料与清漆	用途	结果
喷漆（上色颜料，带胶粘剂与推进器）	可以应用于大多数的材料。然而必须注意的是在这种溶剂与材料之间不会发生化学反应（例如，它与聚苯乙烯泡沫板是兼容的）	可以使用不同种类的喷漆（丝质亚光或高光）来匹配材料表面的色彩
清漆	可以应用于多种不同的材料。形成清澈、透明的涂层，不需要使用额外的材料，就可以创造出细微的差别	例如，可以带来基地、地形与建筑的细微差别，改变材料表面色彩的饱和度与亮度。清漆还可以保护材料的表面（例如，用于芬兰板，组织材料变黄）

续表

丙烯酸树脂漆	用于给模型中特殊的建筑构件或表面涂漆。它易于混合，而且可以用水洗净刷子	通过涂上统一的单色涂层，不同的材料可以相互匹配；非常浓，亮色（取决于色彩沉淀的程度）
木器漆	这种漆可以用来改变木材表面的天然色彩。而木材的原始属性，例如纹理，则被保留下来	通过改变色调，可以创造出稍有不同的色彩与细微的差别。有色木器可以使材料的外观有很大的不同
油与蜡	木材与卡纸板的表面可以使用这些产品进行处理，从而改变色彩与亮度	这些保护性的表面涂料通常可以用来加强木材表面属性的效果

P65

石灰、黏土与模型黏土

石灰（石膏）与黏土，就像在这里所提及的其他材料一样，构成了一个独特的材料类别。人们通常会有局限性，只选择使用那些能够制作整个模型的材料。这些材料尤其适用于不是关注于特定形式细节而是关注于建筑形体的形成与体量的那些表现中。石灰，可以加入其中，它可以被铸成非常精确、光滑的形式。

石灰

石膏被用作1:1建筑的建造材料（例如，以石灰与腻子的形式出现）。它的化学名称是硫酸钙，是一种自然的化合物，也是发电厂的副产品。

在模型制作的过程中，它被用作一种浇筑化合物。首先必须制作模具。这是一个耗费时间的过程，但是它的优点在于你可以按需要经常地浇筑产品。

石膏，作为"模型制作石灰"是白色粉末状。它可以与水混合以液态的形式使用，也可以加工成黏稠状用灰刀进行操作。

液态的石灰会很快地凝固与变硬。一旦它变成这种形态，就可以用砂纸打磨、刷涂涂料与清漆了。

黏土

很久以来，黏土就被用在不同的领域制作多种多样的产品。

建筑模型的历史表明，自从第一次制作模型开始黏土（也被称为肥土）就已经被用于特定领域了：

— 用于构建某种形式：自由形式、有机形式的物体；

— 雕塑；

— 三维实体模型，例如，用于城市发展方案。

黏土的最大优点在于它的使用简单。对于黏土的加工也是一种有触感的体验。使用黏土你可以创造出"有形"的三维草图，影响整个发展过程，并且迅速与不费力气地进行修改。当你操作黏土的时

图 68：
城市设计模型，用陶土制成

候，要确保黏土总是具有足够的湿度（用水）以防止它干燥与碎裂。

黏土模型准备好以后，必须要在窑中进行烧制以保持它的形状。烧制的产物被称为陶器或瓷器。它们用于建筑工业中，用来制作砖与黏土屋面瓦。

塑性模型黏土

除了经典的材料石灰与黏土外，还有其他的产品：

— 风干模型黏土：它的加工性能与黏土相似，不同之处在于它在空气中就可以硬化，并且可以立即用打磨工具进行加工——类似于木材。
— 塑性模型黏土/橡皮泥：这种模型制作材料本身就保持着可加工与可成型的属性，同时也具有稳定的硬度。这种材料的典型特征是在较高的温度下它会变得“相对柔软”，但是在普通室温的情况下，它仍能够重新获得相对结实的坚固性。橡皮泥是一种熟悉的灰绿色。塑性模型黏土有许多不同的颜色。

装配构件越来越经常地用于模型制作当中。

P67

配景：树、人与车

当业主第一次走进一座新建筑的时候，他们通常会惊呼：“它比我想像的小（或者大）得多。”这并不奇怪，因为在建筑建成之前他们对于方案尺度的惟一印象是来自图纸与模型。而这就是问题的症结所在：建筑师怎样才能正确地传达出最终结构的尺寸与尺度的印象呢？为了给观赏者以正确尺度的概念，他们可以使用“熟知的量度”。这种参考点给观赏者创造了一个基准，依据基准他或她可以想像最终的产物。在那些能够传达尺度概念的物体当中有小尺度的人类形象。我们对于自己的高度都有概念，并且我们还可以将这种概念同模型联系在一起。如果我们以这种“人类的尺度”为参照来观察并理解物体的话，会发现对于它们的判断会变得更加容易。

例子：

模型人是一种模具生成的产品，具有所有的尺度。在常用的比例 1:200 的情况下，可以将谷物或稻米垂直地粘在基地上，作为人的一种抽象表现。

配景

除了人，还有许多其他的物体可以帮助我们想像方案的尺度。交通工具在专业经销商那里可以获得，或者通过其他的物体表现，在多层停车场中是必不可少的。这种情况也同样适用于树木与植物，如果它们是设计理念（在景观规划或城市设计方案中）的一个重要特征，那么必须将这些树木表现出来。球体、杆状物以及简单的三维体块都可以用于表现树木。然而最好的方法则是直接使用树木本身。各种干燥的植物（例如西洋蓍草，它通常用于花卉布置）都是用于制作微缩树木的理想材料。如果树木在模型设计（基地设计、建筑与树木以及内部庭院的关系，等等）中扮演了重要的角色，则必须十分小心地选择抽象的树木，因为它们在塑造总体印象时通常起到决定性的作用。

图 69：
表现树木所使用的材料

图 70：
木球

图 71：
树枝/枝条

图 72：
泡沫

图 73：
泡沫

图 74：
高阻泡沫

图 75：
毛刷树

在建筑模型中，以上所提及的各种配景通常已经是你的全部所需了。然而在某些情况下，其他的物品或许更加适于表明未来建筑的意图：

— 街灯作为排列成线的构件；
— 街道家具（长椅、广告柱、标识）；
— 在室内（例如商场与餐馆）中的家具；
— 火车、轮船与飞机，用在交通建筑中。

总结：很明显，以上所提及的大多数材料并不是为了制作模型而专门生产出来的。它取决于你，作为模型的制作者，你应该运用你的想像力将普通的材料应用于建筑模型的语境中去。

提示：

当你选则合适的材料的时候，要考虑一下你将使用或保留这个模型多长时间。许多材料变旧——在这方面模型与建造出来的建筑是非常相像的——会随着时间的消逝而改变着模型的外观。阳光（紫外线辐射），尤其会使卡纸板与木材颜色变深。例如芬兰板，则会很快变黄。虽然时间会影响木制模型的美丽外观，但它对卡纸板模型的耐久性则会有更坏的影响。

P70

从图纸到模型——步骤与方法

接下来，我们将检验制作模型的各种不同方法，并且告诉你可以自己制作模型的不同途经。这些小贴士意在成为帮助你进行创造性工作的指南。

P70

一些初步的想法

展示的目的

在你拿起你的裁纸刀与切割尺之前，你必须思考你的目标。虽然模型制作可以是非常富于创造性的建筑设计方法，而且也会是一个令人非常愉快的活动（例如，将纸片撕开并粘结在一起），但是如果你是为了展示的目的想要制作一个模型的话，在一开始你就必须避免那些实验性的与错误性的操作。因此你必须明确你的目的才能开始。

你必须首先决定你要表现什么。如果这是一个基本的设计原则（例如，一座有着石板与玻璃的会堂），那么它将会影响到你对比例、材料以及抽象程度的选择。如果你想要将一个概念展示给例如学术委员会或者房屋业主的代表，你必须明确地决定你要与目标人群交流些什么。你也必须考虑他们是否可能具有很好的想像力。一般来说，教建筑学的人在鉴赏设计方面有着更多的经验，而且比起那些在建造领域没有什么经验的建筑委托人来说，他们可能更熟悉抽象的模型。

方案与模型的关系

此外，用于展示的方案图也会影响到模型底板的材料与形式的选择，尤其是如果方案图与模型想要形成一个统一的整体的时候。因此你必须仔细地选择那些关键性的材料，并且在图面上探究不同的抽象程度——如果必要的话，也会用到草图。

决定方案的细部

在你的模型中表现出来的细节的程度将取决于你要使用的底板的空间关系与物理尺寸。如果方案中的一些部分（例如，在周边环境中显著的轴线）是非常重要的，但是却又因为过大而不能够放到合适的基地中，你或许就不得不重新考虑你的模型的比例了。

P70

底板

如果周边环境与地形在你的理念中扮演了决定性的角色，那么你应该一开始就思考底板的做法。虽然这种方法或许并不适用于每一个方案，但却总是一个好的开端。以制作场所的工作模型作为开始，当然这种方法同样也可以用于表现模型。一般来说，当人们开始做一个设计时，他们会缺少想法，所以再现地形与地貌并对基地进行空间上的分析通常是值得的。最经常发生的情况是，当你开始工作时，最好的想法出现了。在一开始就制作地形的另外一个原因是越到方案快结束的时候，工作量通常会越多。因此在你做其他任何事情之前，准备并完成这一部分工作是十分必要的。

当制作地形（“基础”）的时候，要确定整个模型的尺度是正确的：如果模型过大，那么它将会非常重而且很难运输；如果模型过小，那么在展示的过程中它将不能够传达出足够的信息。用木质材料制成的板通常是很好的基础（见“材料”一章）。

如果你已经有了一块合适尺寸的底板用于表现你已经选定的部分，那么现在你需要的则是一张总平面图，以将周边环境或者地形反映在底板上。一旦你开始做这项工作了，你必须要做的将是把你的平面的图形信息转化成模型制作材料。你可以运用下列方法中的一种：

— 拓印法：将图纸复制成镜像图形，用胶带将其粘贴在材料上，接下来在丙酮溶剂的帮助下将图形拓到材料上。在这个加工过程中，打印墨水从图纸转移到了材料上。

— 用针来描摹图纸：将图纸放置在材料上，然后用针或者印花粉齿轮（一种专门的模型制作工具）将关键的特征刻在材料上，接下来再描摹那些几何线性图案。

— 绘制：如果只是制作简单的轮廓，最好的方法是用细铅笔在材料上画出所需要的线。

小贴士：

将地形的轮廓转移到模型上通常是一件非常耗时的事。如果你要制作浮雕模型，就需要选择一块材料，其厚度在所期望的比例上要与地形平面上的标高线相符合。这种方法放弃了不必要的进一步的修改。首先要制作所需数量的标高层板，标高层板要适合底板的尺度，然后将整个一块标高层板粘贴在底板上。现在再将第一个标高层从基地平面的基础图上切割出来，基地平面图也被切割成了同样的尺寸，然后再将它转移到下一层纸板。当你已经粘上了第一个标高层以后，就要切割掉额外的部分，一次移动一个标高层到前一个层上，直至你创造出来一个完美的地形的复制品。注意你或许将会不得不制作一个新的建筑基地的地形模型，以便将这个新地形与浮雕模型嵌合在一起。

P72

制作单独的建筑构件

制作模型更像是建造一座预制的房屋，而不是矗立起全部的建筑（从基本的结构到完成的产品）。一般来说，可取的方法是先将建筑构件完全制作好，然后再将他们与其他的构件连结在一起。典型的构件包括：

— 楼板、顶棚、屋顶以及结构当中所有的水平建筑构件；

— 墙、剪力墙以及立面，包括窗与门洞口的布置以及全部其他的垂直构件；

— 室内的墙。

从平面图与草图开始的制作

这些构件是在建筑图纸的基础上制作出来的（与建筑在建筑规划或工作草图的基础上建造出来使用的是同样的方法）。我们在纸面上所拥有的全部都是标准的二维方案图纸，例如，平面图、剖面图与立面图。而这些平面图与剖面图是同样重要的。它们中的每一个都提供了关于建筑的二维视图。然而，放在一起，它们就会造出一个三维的图景。最好的方法就是将这些平面图形缩放到与模型相同的比例，然后将它们转移到材料上。有许多不同的方法可以用来进行这步操作：

— 墙与承重构件依据平面图的示意进行拼装（三维的拼装）。无论你是在工作模型还是城市设计模型上进行操作，这都是一种快捷与简便的方法。

— 将平面上所有的重要信息都转移到要用的材料上（见上文）。

平面中的信息一旦经过确认并且体现在一层又一层的模型中，那些在纵剖面与横剖面中的高度特征也就会逐渐地呈现出其具体的形态了。

小贴士：

当你将构件的长度与高度移植到模型上时，一定要注意一下材料的厚度。如果你想要将墙角斜拼连接起来的话，制作的时候它们的每一条边都必须是完整的长度。然而如果构件要对接在一起的话，你就必须将伸出的那一个构件去掉材料的厚度。

图 76：
将“楼梯板”切割成正确的宽度

墙通常是建筑当中最重要的部分。当你制作模型的时候，你不要将墙作为单独的构件来进行设计，而是要将其作为整体的一部分，对它们作相应的处理。在这方面，模型制造者的工作可以比建造者的工作有效得多，因为他们可以将墙切割或锯成无限长的板条，而当需要它的时候再将它们切割成需要的长度。

这种“无限加工”的方法极大地简化了模型制作工序，它既可以用于城市建设项目（包括一排排有着相同横剖面的房屋或建筑，可以做成“无限长的一排”），也可以用于有着相同横剖面的其他建筑构件（承重构件、窗框、楼板，等等）。制作大量的单独的构件然后再改变它们是值得的，尤其是在你使用模型进行试验的时候。

楼梯

通常楼梯在任何方案中都是决定性的构件，而它们的表现是非常重要的。在工作模型中，楼梯被简化而制成倾斜的表面形式（坡道）。这项工作可以很快完成，而每个人都知道这种做法随即产生的倾斜理应代表什么。用踏面板与踢面板来表现真实的楼梯会多花费一些时间，但是它却能创造出更加打动人的效果。

例子：

当你制作楼梯的时候，要考虑到你的模型中要有多少个台阶。通过制作很宽的梯段（在只有单独梯段的楼梯的情况下），然后用裁纸刀或锯切割出所需要的单独的部分，你可以给自己节省很多工作量。

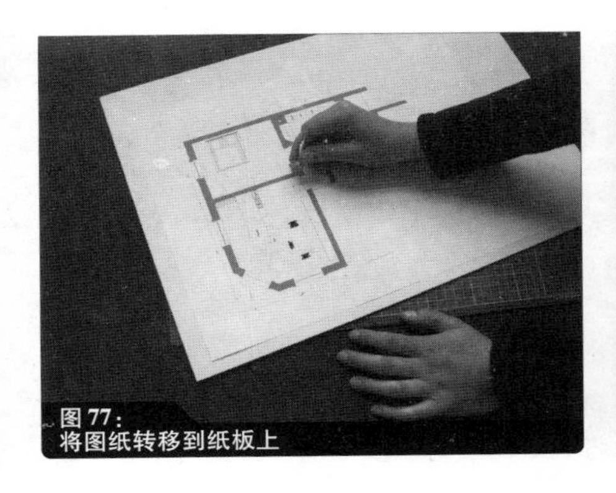
图 77：
将图纸转移到纸板上

图 78：
切割出底板

图 79：
在条板上量出墙的尺寸

图 80：
切割出洞口

图 81：
将墙体粘在一起

图 82：
完成的模型

P75

组装构件

这些建筑构件也应该有条理地组合在一起，因为在大多数情况下，一旦它们粘在一起后你就不可能再将它们分开而不引起任何损坏。一个有效的方法是，首先安装建筑的两片外墙，然后接下来粘结室内的墙与顶棚。最后你可以将遗漏的室外墙体粘贴上，完成整个建筑。

精细模拟的窗子与立面构件对于塑造建筑模型的表现力来说也是十分重要的。你应该先安装窗户，然后再组装室外墙体与立面。对于顶棚上为楼梯开的洞口以及带洞口的墙与屋顶来说，也要使用相同的工序。

有时候，预制全部的与某个特定的制作阶段相关的建筑构件——也就是整个建筑的全部构件——然后再将全部的单元组装在一起是有意义的。然而你应该首先确定，如果有需要，你可以对已经连接在一起的构件进行调整，以防建筑的各部分在几何尺寸上不能够精确地组合在一起。

P75

最后的工作与配景

当所有的构件都已经安装完毕组成了一座建筑以后，还需要有一些最后的工作才能完成整个模型。这个过程必须仔细规划，尤其是如果你要给个别的建筑构件或者整个模型刷油漆的时候。如果，例如，有一片穿过整个建筑的体量是某种确定的颜色，你应该将这个体量制作出来并将它们粘在一起组成一个单元，这样你就可以给这个部分刷

小贴士：

你必须避免在已经粘在一起的模型上使用裁纸刀，因为你不可能像在切割垫板上那样进行准确的操作。更好的方法是尽可能多的制作预制构件，然后再将它们粘结起来。你可以将各部分临时连接起来以检查测量错误。通过以正确的角度将局部竖立起来，你可以确认模型的稳定性。

涂颜色而不会影响到模型的其他部分，甚至如果必要的话还可以使用喷枪。如果模型中的几个部分是确定的颜色，你也应该考虑是在刷涂颜色之前还是之后是添加配景（如树与人）的最佳时机。

添加配景

在模型临近完工的时候，考虑是否并且如何使用配景是绝对必要的。树是一种十分受欢迎的元素，它可以充分地烘托建筑模型。即便如此，树木与其他的配景也只能在它们与你的概念相关时才能加入进来，如果它们仅仅是装饰那就没有必要了。通常，模型的制作者错误地“种植”了过多的树在模型的底板上，这样做会分散人们对模型真正意图的注意力。一般的原则是如果树木是建筑元素的话——这就是说，如果它们具有塑造空间的功能——它们就应该是表现的一部分，但是它们并不是绝对必须的（在一般情况下少就是多）。›✎

P76

表现

完成的模型通常是展览的一部分，或者同语言结合在一起进行方案展示。你一定不能将这最后一步交给偶然来控制。当在墙上悬挂图片时，图片的位置要在视平线上以尽可能获得最佳的效果。模型也应

✎

小贴士：

如果配景扮演着尤其突出的角色，它们将会削弱模型的本质特征：因为有了树木，观赏者将不会真正地去欣赏建筑。为了将这种影响降至最小，你可以将选定的材料涂上颜色以与其他所使用的材料相匹配。不仅是那些构件将会与环境融合得更好，而且仍然可以以空间的逻辑去欣赏它们。如果你要将大型公共建筑例如音乐厅、博物馆以及其他类似的建筑制作成小比例的模型，你可以加一些小的抽象的人来指示通道、强调公共空间并且使观赏者更好的理解真实的逻辑关系。如果你想创造一种氛围，使结构散发出吸引力，你或许可以考虑在主入口处添加大量集中的成组人群，而分散的穿越场所的人要少一些。

该用这种方法来展示，要使观众从良好的视角来观赏它们。你应该保证当人们欣赏一个室内模型的时候，可以看到空间内部而不用采取扭曲的或不舒服的姿势，而城市设计模型则是从上方来观看的。

建造基座

为了保证模型在正确的高度上被展示，你可以制作一个基底（例如以底座的形式）与模型的形式相符合。基座可以用于在概念上支撑模型：例如，如果展示一座高的狭窄的建筑，高的狭窄基座在远处就可以说明建筑的概念。

组群模型的基座通常安装以轮子，这样他就可以从一个展览到另一个展览之间轻松地移动。在一个新的会场上你需要做的全部事情就是把插接构件放回原处。

陈列柜

高质量的模型可以在展柜中进行展示，既可以保护它们的表面也可以强调建筑表现类似于观赏对象的特征。一个有机玻璃的展柜在展览中可以保护模型免受损坏，在公开展览中尤其有用。

将模型挂在墙上

城市设计模型（景观的浮雕模型）可以像一幅画或一张平面图（一种三维的方案图）一样挂在墙上。在这里，预先知道会是什么类型的墙是很重要的。如果方案要挂在隔断上，它们就不能像普通模型一样重。

小贴士：

你或许会不愿扔掉你的模型，但是在你学习期间，你将要做许多的模型，你会没有足够的空间来将它们都保存下来。因为如果将模型保存在地下室或阁楼中，它们会因潮湿或温度的升降而损坏，你可考虑在展示后将你的模型挂在墙上。用这种方法，你既可以保存它们，也可以用这些模型来装饰墙面。

P78

总结

建筑学的学习使学生面临着过多的要求与挑战。在所有的其他事情中，他们还必须学习制作高质量的模型。这些模型当中有许多是为了展示目的而创作的，而且是赶在教学任务就要结束时完成的。这些模型在设计与制作工具方面表现出的潜能被忽视了。学习建筑学包括学习技术问题以及艺术和创造性的方法。模型是一种探讨空间与比例并且增加三维想像能力的方法。模型能够显示出与直面草图之间的不同，也能够帮助学生培养依据二维图纸想像空间关系的能力。

模型也是与非专业观众之间进行交流的方法。如果委托人雇用一名建筑师来设计房屋，最终是模型而不是细致的平面图会使他或她对于建筑师的“产品”有一个具体的概念。模型提供了一个基础，用以推进或者抛弃一些理念，取得进展并且充实细节。对于建筑师与委托人来说，不同的细致程度所扮演的不同角色都反映在他们回应模型的不同方法中：委托人将模型视作完成的结果并且要求高度的精细，因为它们急于看到项目的完成。另一方面，建筑师希望模型是适度的准确，因为这样将会给他们以更大的创造自由。

对于以上两种情况来说，模型都提供了用以表现方案的方法。作为设计理念的三维体现，模型完全与表现图纸同样重要。

P79

附录

P79

致谢

特别感谢：

— 建筑理论与设计系主席，阿诺·雷德勒（Arno Lederer）教授/丹尼尔·马尔奎斯（Daniele Marques）教授，卡尔斯鲁厄大学（University of Karlsruhe）

（照片中的模型是在设计课最后几年以及设计理论课最后一个学期的设计中制作出来的。熟练的模型制作者曼弗雷德·纽比格（Manfred Neubig）开设了模型制作课来配合设计进程）

— 格斯泰克尔－建筑股份有限公司（Gerstäcker-Bauwerk GmbH），模型制作材料与艺术供给，托马斯·路德（Thomas Rüde）与马克·施勒格尔（Marc Schlegel），卡尔斯鲁厄

（在给工具与模型制作材料拍照的时候，提供了关于材料的建议与帮助）

— 建筑系的木工车间，车间指导沃尔夫冈·斯坦因海普（Wolfgang Steinhiper），卡尔斯鲁厄大学（协助拍照）

— 建筑系的金属车间，车间指导安德烈·海尔（Andreas Heil），卡尔斯鲁厄大学（协助拍照）

— 克里斯多佛·鲍曼（Christoph Baumann），卡尔斯鲁厄（模型照片）

— 韦雷纳·赫恩（Verena Horn），卡尔斯鲁厄，（协助拍照）

— 彼德·科比斯（Peter Krebs），建筑学教研室（Büro Für Architektur），卡尔斯鲁厄（模型照片）

— 史蒂芬尼·施密特（Stefanie Schmitt），施图登湖（Stutensee）（模型照片）

P80

图片出处

以下的照片是由卡尔斯鲁厄大学建筑理论与设计系主席阿诺·雷德勒教授/丹尼尔·马尔奎斯教授慷慨提供的。雷德勒教授指导设计。研究助理有克里斯汀·巴贝（Kristin Barbey）、罗兰·科茨（Roland Kötz）、彼得·科比斯（Peter Krebs）与贝吉特·梅尔霍恩（Birgit Mehlhorn）。曼弗雷德·纽比格监督与设计过程并行的模型制作。如果没

有另外说明，照片都是由西洛·梅肖（Thilo Mechau）拍摄的，他是建筑系摄影工作室的职员。

图 1　学生方案，设计，雷克雅未克港，弗洛兰·博姆勒（Florian Bäumler）

图 5　学生方案，设计，共和宫（Palace of Republic），鲁文·莱姆普恩（Ruwen Rimpern）城市周边环境模型：曼弗雷德·纽比格

图 6　学生方案，设计，雷克雅未克港，合作完成的城市设计模型。

图 21　第一行：学生方案，设计，"再做一遍"，斯特芬·乌尔茨巴赫（Steffen Wurzbacher）
第三行：学生方案，设计，"再做一遍"，马蒂亚斯·莱赫贝格（Matthias Rehberg）
第四行：学生方案，设计，"再做一遍"，伊万娜·萨雷斯诺（Ioana Thalassinon）

图 24　建筑理论课期末设计方案，合作完成的橡皮泥模型，拍照：科尼留斯·博伊（Cornelius Boy）

图 26　左：学生方案，设计，共和宫，马克·纽丁（Marc Nuding）
右：学生方案，设计，伊斯兰社区中心，阿克塞尔·鲍登迪斯提（Axel Baudendistel）

图 27　硕士方案，巴塞罗那海滨度假区，菲利普·洛佩尔（Philip Loeper）

图 48　学生方案，设计，雷克雅未克港，霍尔格·莱特格鲁特（Holger Rittgerott）

图 49　学生方案，设计，"再做一遍"，山口丽莎（Lisa Yamaguchi）

图 50　学生方案，设计，"再做一遍"，安德烈·乔德（Andrea Jörder）

图 51　学生方案，设计，"再做一遍"，山口丽莎

图 60　学生方案，设计，"再做一遍"，本杰明·弗尔曼（Benjamin Fuhrmann）

图 61　学生方案，设计，雷克雅未克港，马库斯·施瓦泽贝切（Markus Schwarzbach）

图 62　学生方案，设计，雷克雅未克港，库诺·贝克（Kuno Becker）、迪纳·普福特（Tina Puffert）、霍尔格·莱特格鲁特（Holger Rittgerott）

图 66　学生方案，设计，共和宫，马蒂亚斯·莫尔（Matthias Moll）

图 67　学生方案，设计，雷克雅未克港，合作完成的模型

图 68　学生方案，设计，伊斯兰社区中心，阿克塞尔·鲍登迪斯提（Axel Baudendistel）

以下的照片是由卡尔斯鲁厄建筑学教研室的彼德·科比斯提供的。

图 4　竞赛模型，梅因哈特（Mainhardt）社区中心

图 12、13、43 竞赛模型，圣奥古斯丁教堂（Church of St Augustine），海尔布隆（Heilbronn）

图 14　竞赛模型，斯特拉·玛丽斯礼拜堂（Stella Maris Chapel），斯图加特

图 25　竞赛模型，圣乔治教区会堂（St George′s parish hall），里德林根（Riedlingen）

图 65　竞赛模型，社区中心，施韦青根（Schwetzingen）

以下的照片是由卡尔斯鲁厄的德克里斯多佛·鲍曼提供的：

图 70、71、72、73、75 研究模型中的细部

以下的照片是由多特蒙德的伯特·比勒费尔德（Bert Bielefeld）提供的：

图 11、16、45－57、74

以下的照片是由多特蒙德的伊莎贝拉·斯奇帕（Isabella Skiba）提供的：

图 8－10、44

以下的照片是由施图登湖的史蒂芬尼·施密特提供的：

图 18　研究模型，圣弗朗西斯幼儿园（St Francis Kindergarten）

图 19　第二行：硕士项目，奥林匹亚站（Olympia station），斯图加特

图 63　硕士项目，奥林匹亚站，斯图加特

作者提供了图 2、15、17、20－23、28－42、52－59、64、69、76－82